Handbook of Nanomagnetism

Handbook of
Nanomagnetism
Applications and Tools

edited by

Rosa A. Lukaszew

Pan Stanford Publishing

Published by

Pan Stanford Publishing Pte. Ltd.
Penthouse Level, Suntec Tower 3
8 Temasek Boulevard
Singapore 038988

Email: editorial@panstanford.com
Web: www.panstanford.com

British Library Cataloguing-in-Publication Data
A catalogue record for this book is available from the British Library.

Handbook of Nanomagnetism: Applications and Tools

ISBN 978-981-4613-04-0 (Hardcover)
ISBN 978-981-4613-05-7 (eBook)

Printed in the USA

Contents

Preface

Nanomagnetism comprises magnetic phenomena in matter where at least one dimension is in the nanoscale, i.e., in the range from 1 to 100 nm. This field includes the study of the properties and applications of magnetism in nanoparticles, nanodots, nanowires, thin films, and multilayers, and also macroscopic samples that contain nanoscopic particles. Applications of nanomagnetism range from spintronics—i.e., manipulation, storage, and transfer of information by means of electron spins—to biosensors as well as medical applications.

Nanomagnetism has been able to stimulate applied as well as fundamental research, particularly in magneto-transport as well as magneto-optical properties of nano-size materials. In particular, spin tunneling properties are exploited in spintronic applications and are a unique tool for surface imaging as in spin-polarized scanning tunneling microscopy.

Thus, this book compiles the work of leading researchers specializing in spintronics, particularly spin transfer torque (STT) and its application to next-generation MRAM (STT-RAM); in highly ordered (e.g., $L1_0$) magnetic alloys in thin film form; and in perpendicular magneto-recording media and STT junctions, as well as relaxation mechanisms in magnetic materials for spintronic applications.

The book also includes chapters on nonlinear dynamic properties of 2D arrays of magnetic nano-dots, dynamics of magnetic nanoparticles and nanodevices, as well as nonlinear optics, and magnetoplasmonic studies in magnetic nanostructures with an eye to biosensing applications.

The last two chapters are more fundamental in nature and are related to understanding magnetic exchange in molecule-based magnets, and spin-polarized scanning tunneling microscopy, a sophisticated technique that enables studying magnetic phenomena on surfaces with real-space imaging and spectroscopy techniques down to the atomic level.

Rosa Lukaszew
June 2015

Chapter 1

Introduction

Rosa Alejandra Lukaszew
Physics Department, College of William and Mary, Williamsburg, VA 23185, USA
ralukaszew@wm.edu

1.1 Overview

Remarkable progress has been achieved in recent times toward the development of technological applications of nanomagnetism, stimulated by the small gap that exists between basic studies and applications. Thus, this book concentrates on recent developments of interest at both application and fundamental fronts, as described by leading experts in nanomagnetism. The range of topics covered spans *spintronics* applications; chemically ordered microstructures leading to strong magnetic anisotropy such as $L1_0$ alloy thin films and nanostructures; relaxation dynamics in spintronic materials, magnetic nanoparticles, and nanodevices; magnetic exchange in molecule-based magnets; nonlinear dynamic properties of two-dimensional arrays of magnetic nanodots; nonlinear optics in magnetic nanostructures; and magnetoplasmonic materials and spin-polarized scanning tunneling microscopy (STM).

Handbook of Nanomagnetism: Applications and Tools
Edited by Rosa A. Lukaszew

ISBN 978-981-4613-04-0 (Hardcover), 978-981-4613-05-7 (eBook)
www.panstanford.com

It is useful to define what magnetism is and how it is relevant to the magnetic properties of materials before discussing the nanoscale influence on the magnetic properties. Magnetism arises from the properties of electrons in materials. Electrons have an intrinsic angular moment or spin (**S**) and an orbital angular moment (**L**). Thus, the electron has a magnetic dipole moment (μ) associated with its intrinsic spin, such that

$$\mu_s = -(e/m)\mathbf{S} \tag{1.1}$$

where e is the electron charge, 1.6×10^{-19} C, and m is the electron mass, 9.11×10^{-31} kg. The minus sign indicates that **S** and μ have opposite orientation. It is important to note that neither **S** nor μ can be measured directly; only one component along any axis can be measured. Also, since **S** is quantized, it follows that $\mathbf{S}_z = m_s(h/2\pi)$, with $m_s = +/-1/2$ and h is Plank's constant, so μ is

$$\begin{aligned}\mu_{s,z} &= -(e/m)\mathbf{S}_z = -(e/m)m_s(h/2\pi) = \pm eh/4\pi m \\ &= 9.27 \times 10^{-24} \text{J/T} = \mu_B \end{aligned} \tag{1.2}$$

where μ_B is the Bohr magneton. Similarly, the magnetic moment associated with the electron's orbital magnetic moment (**L**) is

$$\mu_{orb} = -(e/2m)\mathbf{L} \tag{1.3}$$

Spin-dependent electron interactions constitute the primary basis for describing magnetism. In a solid these interactions require descriptions of electronic overlap within the context of both real space and binding energy space; thus, the magnetic properties of materials arise from magnetic moment interactions, where magnitude, sign, and strength of the coupling and the range over which the resultant magnetic order extends qualify such interactions.

1.2 Magnetic Order

The magnetic characteristics of atoms depend on the arrangement of electrons and the magnetic moment associated with the electrons spin. Thus, atoms possess an overall magnetic moment when there are unpaired electrons in an orbital. The types of interactions can be summarized as *exchange interaction*, which is electrostatic in nature; *dipolar interaction*, related to spin–spin coupling; and *anisotropy interaction*, related to spin–orbit coupling. It is worth noting that exchange coupling is much greater than dipolar coupling for nearest neighbors but is short ranged since it falls off exponentially compared to dipolar coupling that falls off as $1/r^3$. Within the exchange interaction we distinguish *direct exchange* in which the magnetic ions interact because their charge distributions overlap, *superexchange* in which magnetic ions with nonoverlapping charge distributions interact because of overlap with the same nonmagnetic ion, and *indirect exchange* in which in the absence of overlap, a magnetic interaction is mediated by interactions with the conduction electrons.

In solids, electron–electron interactions between atoms can lead to *magnetic ordering*. Thus, below some critical temperature depending on the particular material (T_c) and when there are significant magnetic interactions magnetically ordered states can exist, even with no applied field, such as *ferromagnetic interactions* where all local moments have a positive component along the direction of the spontaneous magnetization, *antiferromagnetic interactions* where individual local moments sum to zero total moment and there is no spontaneous magnetization, and *ferrimagnetic interactions* where local moments are not all oriented in the same direction, but there is a nonzero spontaneous magnetization. Also, in bulk samples, dipolar coupling can alter spin configurations favored by short-range exchange coupling and the overall magnetic energy is minimized by the formation of domains. Upon domain formation, the dipolar energy is minimized and the exchange energy is only raised for a small number of sites at the domain boundaries. When domains form, the domain wall thickness is dictated by a competition between exchange and anisotropy energies. In small fields, the domains can reversibly align with fields by smooth motion of domain walls (Fig. 1.1). At high fields, the domains align irreversibly with the fields since the material has defects and the defects can prevent domain walls from returning to the original zero bulk magnetization. Thus, a reverse field is required to return to zero

Figure 1.1 Magnetic force microscopy (MFM) image showing the domain configuration for an 80 nm $L1_0$ FePd thin film with perpendicular magnetic anisotropy. The bright and dark regions correspond to domains with magnetization pointing in opposite directions, in this case perpendicular to the surface of the thin film. The image is 3 × 3 μm in size.

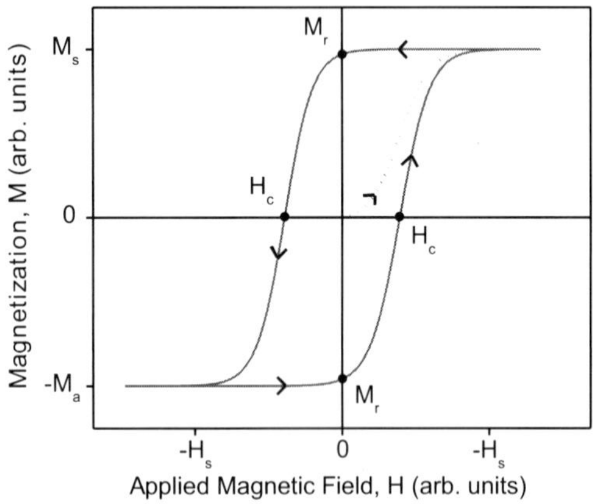

Figure 1.2 Typical symmetric hysteresis loop for a ferromagnetic material. Upon saturation (H_S) the magnetization is highest (M_S). The materials exhibit remnant magnetization (M_r) when the applied field is reduced from saturation to zero. The field required to reverse the magnetization direction is the coercive field ($\mathbf{H}_c$).

bulk magnetization (coercive force) and this is responsible for the hysteresis observed in magnetic moment versus applied magnetic field curves (Fig. 1.2). Typical domain walls have a thickness of a few nanometers, reflecting the competition between nonrelativistic exchange and relativistic magnetic anisotropy.

1.3 Nanomaterials and Magnetism

Materials properties change fundamentally when the dimensionality of the system is reduced. Thus, nanoscopic samples exhibit important differences compared to the magnetism of macroscopic samples, arising from the fact that nanoscopic dimensions are comparable to characteristic magnetic lengths, and also from broken translation symmetry, which results in sites with a reduced coordination number, broken exchange bonds, and frustration. In fact when matter is in the nanoscale size its magnetic properties are affected on different length scales, from a few interatomic distances for intrinsic properties such as magnetization and anisotropy to several interatomic distances, $\sim$10 nm, for other properties such as the coercive field. Compared to macroscopic bulk samples, the magnetic interactions in a nanostructure can be significantly enhanced and it is also possible that new interactions that have been forbidden by symmetry in an extended three-dimensional system can now become quite relevant. An example of the effect of characteristic lengths on the magnetic properties are magnetic particles with dimensions smaller than the critical magnetic single-domain diameter, leading to single-domain structures as the configuration with the lowest energy.

Nanoscopic objects also exhibit a higher proportion of surface and interface atoms, which in general are in close contact with other physical systems, for example a substrate or proximity with another material layer in the case of a thin films. Because of this it is possible to modify the magnetic properties. This is particularly noticeable in giant magnetoresistance (GMR), a quantum mechanical effect observed in thin-film structures composed of alternating ferromagnetic and nonmagnetic layers. In such multilayers the ferromagnetic layers are separated by very thin (about 1 nm) nonferromagnetic layers (e.g., Fe/Cr/Fe). At certain thicknesses the magnetic coupling between adjacent ferromagnetic layers becomes antiferromagnetic, making it energetically preferable for the magnetizations of adjacent layers to align in antiparallel configuration. This type of exchange was first proposed by Ruderman and Kittel and later extended by Kasuya and Yosida [1] to give the theory now generally known as the RKKY coupling, Thus, the electrical resistance of the device is normally higher in the antiparallel case. The interlayer spacing in these devices typically corresponds to the second antiferromagnetic peak in the antiferromagnetic–ferromagnetic oscillation in the RKKY coupling. This phenomenon led to the Nobel Prize in Physics in 2007 awarded to Albert Fert and Peter Grünberg and has led to the remarkable growth of magnetorecording applications [2].

The phenomenon of electron tunneling has been known since the early days of quantum mechanics. Spin-dependent tunneling in magnetic tunnel junctions, that is, trilayers where two magnetic layers are now separated by an insulating one, has aroused enormous interest. This is so because the large tunneling magnetoresistance (TMR) observed in such junctions garnered much attention due to possible applications in nonvolatile random access memories (RAMs) and next-generation magnetic field sensors. Similar ideas have been pursued even more recently using spin transfer torque (STT), an effect where the orientation of a magnetic layer in a magnetic tunnel junction can be modified using a spin-polarized current. STT can be used to flip the active elements in magnetic random access memory. STT-RAM has the advantages of lower power consumption and better scalability over conventional magnetoresistive random access memory (MRAM), which uses magnetic fields to flip the active elements.

Deposition of different material overlayers can also affect the magnetic properties (Fig. 1.3). For example, it has been observed that vanadium cap layers over magnetic ones, such as Fe or Co, produces partial perpendicular magnetic anisotropy similar to the effect of magnesium oxide layers (MgO) but with the advantage of low intrinsic magnetization damping and also metallic character attributed to a transient ferromagnetic state. This is of interest in applications where magnetization dynamics, magnetoresistance, and magnetic anisotropy are equally relevant, which is the case in novel STT-RAM applications [3].

In the case where all dimensions are in the nanoscale, fewer domain walls are needed in order to minimize the stray field energy. The extreme limit is represented by single-domain particles. These nanosized objects may be dispersed in a liquid matrix or immersed in solid matrices, where each cluster or particle may feel a strong interaction with its immediate neighborhood. In the

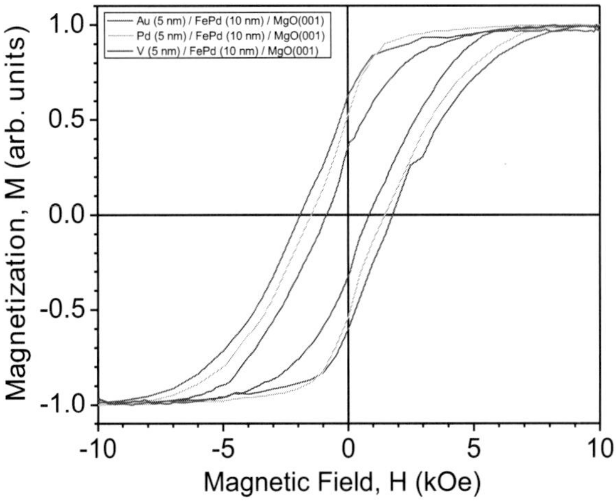

Figure 1.3 Hysteresis loops in perpendicular configuration, illustrating the influence of the overlayer material on the hysteresis loop characteristics for $L1_0$ FePd thin films capped with gold, Pd, or V.

case of nanoparticles, the magnetic moments can be kinetically frozen below the *blocking temperature* (T_B), while they are in *superparamagnetic state* above T_B. This phenomenon is observed when the thermal energy ($k_B T$) is of the same order of magnitude as the anisotropy energy of the particles, leading to an effective zero average magnetic moment.

Recent developments in chemical nanofabrication have enabled fabrication of core–shell nanoparticles or even nano-onions (Fig. 1.4), leading to additional modification of the physical properties of the particles. On this front, it is possible to combine noble metals with ferromagnetic materials and exploit the optical properties of each material leading to new magnetoplasmonic possibilities [4]. Such structures will be further discussed in Chapter 8

Materials where the dimensions are close to the scattering mean free path for various photon–electron interactions, diffusion length in atomic transport, exchange length, etc., can exhibit new properties. Thus, dimensional confinement and large surface-to-volume ratio in magnetic nanostructures can lead to new phenomena compared to bulk, such as exchange bias, magnetic surface anisotropy, GMR, TMR, spin torque, vortex spin states,

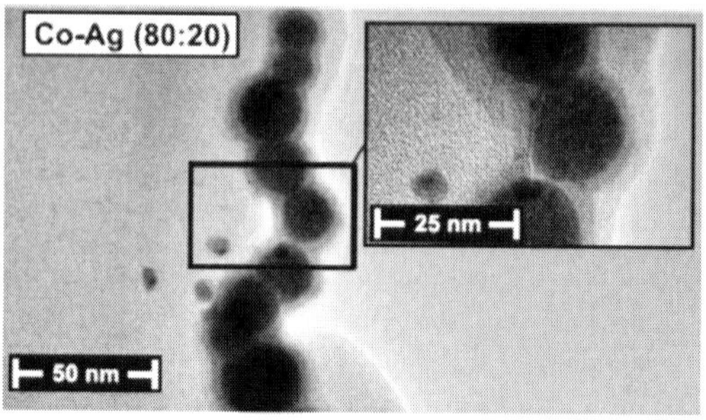

Figure 1.4 Transmission electron microscopy (TEM) image of core–shell CoAg nanoparticles. Here the core is Co and the shell Ag in a, 80:20 volume ratio. The inset shows a high-resolution TEM image clearly revealing the core–shell microstructure of the nanoparticles [4].

superparamagnetism, or magnetostatic modes. The dynamic behavior of nanoscaled objects also differs from that of macroscopic samples of the same materials due to the increased importance of thermal fluctuations. Also, nanostructured materials may present a crystal structure that is different from that observed in bulk samples. This is the case, for example, of metallic cobalt that changes from hexagonal close packed (*hcp*) to face-centered cubic (*fcc*) in epitaxial thin films on *fcc* copper or in nanoparticles with diameters below approximately 30 nm.

Nanostructuring can result in new magnetic states such as superparamagnetism, which, for example, sets a limit for bit size in magnetorecording data applications but is useful in other applications such as magnetic inks and bioapplications. Magnetic behavior in nanoscale systems has inspired a number of new technological developments [5].

Theoretically, a quantitative description of the interactions of spin with its environment can only be obtained on the basis of relativistic quantum-mechanical calculations. Density functional theory (DFT), which is an effective one-particle theory where the output quantities can be described in terms of auxiliary single-particle wave functions that are solutions of Schrödinger-type equation, in conjunction with other models such as Heisenberg and micromagnetic models have been used to extrapolate results into regimes that are difficult to access just with DFT such as very extended length scales.

Since in DFT the orientation of the magnetic moments is independent of the spatial orientation of the crystal lattice, that is, spin space and real space directions are decoupled, in order to introduce coupling between them, the relativistic motion of the electron in the screened potential of the atomic nuclei must be considered [6]. Thus, in the moving rest frame of the electron, the electric field, **E**, is Lorentz-transformed into a magnetic field, **B**, that can couple to the electron's spin, leading to a term in the Hamiltonian that has the form $(\mu_B/2c)\,\sigma(\mathbf{E}(\mathbf{r}) \times \mathbf{p})$ in the Dirac theory of the electron, where **p** is the momentum of the electron and μ_B is the Bohr magneton. It has been pointed out that while the coupling between two spins, S_i and S_j, can be described by a Heisenberg-type term, $J_{ij}(S_i \times S_j)$, in the presence of spin–orbit coupling additional interaction of the spins of the form $D_{ij}(S_i \times S_j)$ coupled to certain symmetry conditions that can occur at surfaces can come into play. Such interaction may not be rare in nanostructures and its reliable calculation has recently been carried out using significant parallel supercomputers resources. Validation for many aspects of these theories has been provided by the development of experimental probes such as spin-polarized STM, which will be discussed in a later chapter [7, 8].

Nanomagetism is an exciting field of research where new predictions and discoveries continue to arise. For example magnetic molecules are a new class of fascinating materials. These molecules contain a finite number of interacting spin centers and thus provide ideal opportunities to study basic concepts of magnetism. Aspects of molecular magnetism are described in Chapter 9. Other materials with promising new magnetic properties are the topological insulators. It has been theoretically predicted that the surface states of topological insulators could provide access to novel, fascinating physical phenomena such as magnetic monopoles or Majorana fermions [9–11]. Recently theoretical studies have shown that breaking gauge symmetry [12] or time reversal symmetry [13] in these materials produces exotic states that may be conducive toward realizing new magnetoelectric effects and tools useful for quantum computing [14]. Further, the quantized version of the anomalous Hall effect has been predicted to occur in magnetic topological insulators and only very recently it has very recently been experimentally observed [15].

References

1. M. A. Ruderman, C. Kittel, Indirect exchange coupling of nuclear magnetic momentsby conduction electrons, *Phys. Rev.*, **96**, 99–102 (1954); T. Kasuya, A theory or ferro- and antiferromagnetism on Sener's model, *Prog. Theor. Phys.*, **16**, 45 (1956); K. Yosida, Magnetic properties of Cu-Mn alloys, *Phys. Rev.*, **106**, 893 (1957).
2. M. N. Baibich, J. M. Broto, A. Fert, F. Nguyen Van Dau, F. Petroff, P. Eitenne, G. Creuzet, A. Friederich, J. Chazelas, Giant magnetoresistance of (001)Fe/(001)Cr magnetic superlattices, *Phys. Rev. Lett.*, **61**, 2472–2475 (1988); G. Binasch, P. Grünberg, F. Saurenbach, W. Zinn, Enhanced magnetoresistance in layered magnetic structures with antiferromagnetic interlayer exchange, *Phys. Rev. B*, **39**, 4828–4830 (1989).
3. K. L. Wang, J. G. Alzate, P. Khalili Amiri, Low-power non-volatile spintronic memory: STT-RAM and beyond, *J. Phys. D: Appl. Phys.*, **46**, 074003 (2013).
4. L. Wang, C. Clavero, Z. Huba, K. J. Carroll, E. E. Carpenter, D. Gu, R. A. Lukaszew, Plasmonics and enhanced magneto-optics in core–shell Co–Ag nanoparticles, *Nano Lett.*, 11(3), 1237–1240 (2011).
5. A. Imre, G. Csaba, L. Ji, A. Orlov, G. H. Bernstein, W. Porod, Majority logic gate for magnetic quantum-dot cellular automata, *Science*, **311**, 205–208 (2006).
6. G. Bihlmayer, M. Heide, Y. Mokrousov, A. Thiess, St. Heinze, St. Blügel, *Anisotropic Magnetic Interactions in Low Dimensions, NIC Symposium 2010*, IAS Series, Vol. 3, 175–182, G. Münster, D. Wolf, M. Kremer, Eds. Forschungszentrum Jülich, Jülich Supercomputing Centre (JSC), John von Neumann Institute for Computing (NIC), Schriften des Forschungszentrums Jülich, ISBN 978-3-89336-606-4.
7. M. Bode, M. Heide, K. V. Bergmann, P. Ferriani, S. Heinze, G. Bihlmayer, A. Kubetzka, O. Pietzsch, S. Blügel, R. Wiesendanger, Chiral magnetic orderat surfaces driven by inversion asymmetry, *Nature*, **447**, 190 (2007).
8. P. Ferriani, K. v. Bergmann, E. Y. Vedmedenko, S. Heinze, M. Bode, M. Heide, G. Bihlmayer, S. Blügel, R. Wiesendanger, Atomic-scale spin spiral with uniquerotational sense: Mn monolayer on W(001), *Phys. Rev. Lett.*, **101**, 027201 (2008).
9. D. Hsieh, Y. Xia, D. Qian, L. Wray, J. H. Dil, F. Meier, J. Osterwalder, L. Patthey, J. G. Checkelsky, N. P. Ong, A. V. Fedorov, H. Lin, A. Bansil, D. Grauer, Y. S. Hor, R. J. Cava, M. Z. Hasan, A tunable topological insulator

in the spin helical Dirac transport regime, *Nature*, **460**(7259), 1101 (2009).

10. X. L. Qi, R. D. Li, J. D. Zang, S. C. Zhang, S. C. Inducing a magnetic monopole with topological surface states, *Science*, **323**(5918), 1184–1187 (2009).
11. J. Linder, Y. Tanaka, T. Yokoyama, A. Sudbo, N. Nagaosa, Unconventional superconductivity on a topological insulator, *Phys. Rev. Lett.*, **104**(6), 067001 (2010).
12. J. Linder, Y. Tanaka, T. Yokoyama, A. Sudbo, N. Nagaosa, Unconventional superconductivity on a topological insulator, *Phys. Rev. Lett.*, **104**, 067001 (2010).
13. L. Bao, W. Wang, N. Meyer, Y. Liu, C. Zhang, K. Wang, P. Ai, F. Xiu, Quantum corrections crossover and ferromagnetism in magnetic topological insulators, *Sci. Rep.*, **3**, 2391 (2013).
14. L. Fu, C. L. Kane, Superconducting proximity effect and Majorana fermions at the surface of a topological insulator, *Phys. Rev. Lett.*, 100, 096407 (2008).
15. C.-Z. Chang, J. Zhang, X. Feng, J. Shen, Z. Zhang, M. Guo, K. Li, Y. Ou, P. Wei, L.-L. Wang, Z.-Q. Ji, Y. Feng, S. Ji, X. Chen, J. Jia, X. Dai, Z. Fang, S-C. Zhang, K. He, Y. Wang, L. Lu, X.-C. Ma, Q.-K. Xue, Experimental observation of the quantum anomalous hall effect in a magnetic topological insulator, *Science*, **340**, 167 (2013).

Chapter 2

Spintronics and Spintronic Applications

Jiwei Lu[a] **and Stuart A. Wolf**[b]

[a] *Department of Materials Science and Engineering, University of Virginia, 395 McCormick Road, Charlottesville, VA 22904, USA*

[b] *Departments of Physics and Materials Science and Engineering, University of Virginia, Charlottesville, VA 22904, USA*

jl5tk@virginia.edu, saw6b@virginia.edu

2.1 Introduction to Spintronics

The term "spintronics" usually refers to the branch of physics concerned with the manipulation, storage, and transfer of information by means of electron spins in addition to or in place of the electron charge as in conventional electronics. Introduced in 1996, spintronics was originally the name for a Defense Advanced Research Projects Agency (DARPA) program managed by S. A. Wolf [1].

Spintronics has inherent merits, including very low power dissipation, nonvolatility, and the possibility of integrating memory and logic into a single device. The control of spin is central as well to efforts to create entirely new ways of computing, such as quantum computing, or analog computing that uses the phases of signals for computations.

Handbook of Nanomagnetism: Applications and Tools
Edited by Rosa A. Lukaszew

ISBN 978-981-4613-04-0 (Hardcover), 978-981-4613-05-7 (eBook)
www.panstanford.com

Modern electronics is charge based where electronic charge is used to encode digital information (binary bits 0 and 1). Charge is a scalar quantity and only has magnitude. Therefore, logic levels must be demarcated by a difference in the *magnitude* of charge. For example, more charge stored in a device could signify the logic bit 0 and less charge could signify the logic bit 1. Or the presence of charge could signify the logic bit 0 and absence could signify the bit 1. This is the approach adopted in the metal–oxide–semiconductor field-effect transistor (MOSFET). When the channel is full of charge, the transistor is on and could encode the bit 0. When the channel is depleted of charge, the transistor is off and could encode the bit 1. Switching between logic levels requires changing the magnitude of charge in the device, which invariably involves current flow and an associated I^2R Joule dissipation (I = current and R = resistance in the path of the current). This dissipation is unavoidable in any charge-based electronics.

Spin, on the other hand, is a *pseudovector* (spinnor) that has a fixed magnitude but a variable direction (or polarization) and whose rotations are governed by the Pauli spin matrices. By placing an electron in a magnetic field, we can make its spin polarization bistable (only polarizations parallel and antiparallel to the field are allowed eigenstates). These two polarizations can encode the logic bits 0 and 1. Switching between logic levels will then require merely flipping the spin, without physically moving the charge in space and causing a current flow. This can potentially reduce energy dissipation significantly since the I^2R loss is eliminated. The energy dissipated during a spin flip (logic switching event) is the energy difference between the two bistable polarization states that are separated by the Zeeman energy. This energy can be made less than the thermal energy k_BT without causing too many unwanted random spin flips and bit errors, because spin couples very weakly with phonons and as a result, spin split levels are not broadened by $\sim k_BT$. If that were not the case, electron spin resonance experiments could never be carried out at room temperature where the spin splitting energy is only few tens of μeV (tens of gigahertz frequency), while the room temperature k_BT is 25 meV. Because of this weak phonon coupling, spins can also be maintained out of equilibrium for long durations. As a result, the energy dissipated during a switching event can be less than the Landauer–Shannon limit of $k_B T \ln 2$ since the latter result is derived on the basis of equilibrium thermodynamics [2].

Spintronic devices are scalable to rather small dimensions and the scaling limit is the superparamagnetic limit, which is the thermal fluctuation limit of about 40 $k_B T$. It translates to a nanomagnet size of about 5 nm^3 for a typical ferromagnet. Below the critical size, the spins in the ferromagnet can randomly flip direction due to thermal fluctuations thus it become superparamagnetic. The requirement of $\sim$40 $k_B T$ is suggested for the information storage for a data retention over 10 years [3]. However, the spin state can be stable for $\sim$1 millisecond even below the superparamagnetism limit, which is long enough to allow some information processes [4].

This chapter describes the fundamental physics of spin-dependent transport, and the topics include the spin polarization, magnetoresistance, and spin torque transfer (STT). A brief description of spintronic devices and the future direction is included.

2.2 Spin Polarization in Ferromagnets

Spin is a fundamental quantum-mechanical property. It is the intrinsic angular momentum of an elementary particle, such as the electron. Thus, any charged object possessing spin also possesses an intrinsic magnetic moment.

Spin polarization spontaneously occurs in ferromagnets. In magnetic materials like iron, nickel, or magnetite, the atomic moments exhibit very strong interactions due to electronic exchange forces and result in a parallel or antiparallel alignment of atomic moments. A qualitative model in explaining the origin of spin polarization and ferromagnetism is called itinerant electron model proposed first by E. C. Stoner in 1933 [6]. It states that the outer electrons in unfilled shells contribute to the magnetic moments of the material. These electrons are not localized and contribute as conduction electrons in the energy band. In the metallic state of 3*d* transition elements, the 3*d* and 4*s* atomic energy levels are broadened into energy bands. In practice, one cannot separate 3*d* and 4*s* orbital since they are strongly hybridized with each other in the solid.

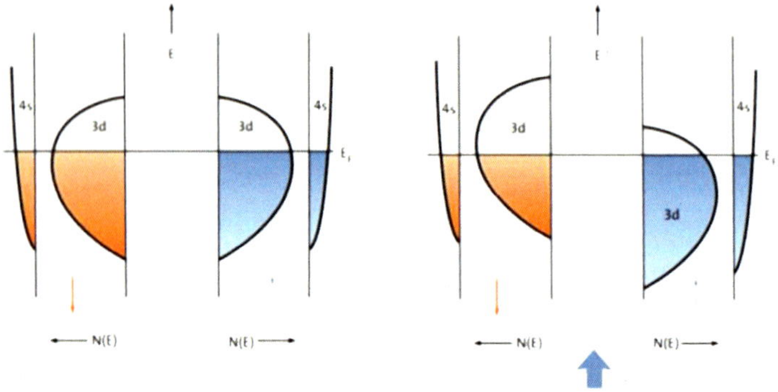

Figure 2.1 (Left) A schematic plot is shown for the energy band structure of a d transition metal. The density of states N(E) is shown separately for the spin-up and spin-down electrons and where a simplified separation has been made between the 4*s* and 3*d* band energies. For the nonmagnetic state these are identical for the two spins. All energy levels below the Fermi energy are occupied states (orange and blue). The colored area (orange + blue) corresponds to the total number of valence electrons in the metal. (Right) The corresponding picture is illustrated for a ferromagnetic state, with a spin polarization chosen to be in the up direction ($N\uparrow > N\downarrow$; blue area > orange area). This polarization is indicated by the thick blue arrow at the bottom figure to the right [5].

According to Pauli's exclusion principle, only one electron can occupy a particular state. However, considering the degenerate spin-up and spin-down states of electrons, two electrons with different spin states could occupy the same state. All energy states are filled up from the ground state up to the highest energy level, which is called Fermi energy (E_{F}). As shown in Fig. 2.1, in a nonmagnetic 3*d* metal there are equal numbers of spin-up and spin-down electrons, resulting in the sum of zero net spin and thus no net magnetization. And thus we deduce that the spin polarization is zero. As a ferromagnetic metal, 3*d* orbitals are partially filled, resulting in unequal numbers of spin-up and spin-down electrons. Here the spin polarization is defined as

$$P = \frac{N_{\uparrow} - N_{\downarrow}}{N_{\uparrow} + N_{\downarrow}} \tag{2.1}$$

where $N_{\uparrow}$ ($N_{\downarrow}$) = number of electrons with spin up (spin-down). Here, $N_{\uparrow}$ is larger than $N_{\downarrow}$, and this leads to a spin polarization $P > 0$ in ferromagnetic metals.

There are two important observations that we need to make here. First is the vertical displacement between the spin-up and spin-down bands due to the exchange energy splitting between the up and down spin densities of states. In this case, more up spins states in blue color are filled than the spin-down states, so the up and down spin are called the majority spin and minority spin respectively. Secondly, we could see that the density of states at the Fermi level $N(E_F)$ are very different for the two spin bands. This would be very important in understanding the spin polarized current and the giant and tunneling magnetoresistance (TMR) phenomenon in the following section.

Mott proposed the two-current model in 1936 [7] when he sought an explanation for an unusual behavior of resistance in ferromagnetic metals. He realized that at sufficiently low temperature, where magnon scattering becomes vanishingly small, electrons of majority and minority spin, with a magnetic moment parallel and antiparallel to the magnetization of a ferromagnet, respectively, do not mix in the scattering processes [8]. The conductivity can then be expressed as the sum of two independent and unequal parts for two different spin orientations. This leads to the conclusion that the current in ferromagnets is spin polarized.

Specifically, as illustrated in Fig. 2.2, the electron transport through the nonmagnetic metal/ferromagnetic/nonmagnetic metal structure, the conductivity is mainly determined by the incident unpolarized 4*s* electrons, which are easily mobile due to the wide energy range of the bands derived for the 4*s* states [7]. Under the two-current mode, we consider the scattering process of the spin-up and spin-down electrons separately when the unpolarized electrons enter the nonmagnetic metal/ferromagnet interface. As we pointed up in the previous section, for ferromagnets with spin up as the majority spin, there are much more spin-down than the spin-up states available at the Fermi level. Therefore the spin-down 4*s* electrons experience a much stronger scattering at the interface compared with the spin-up 4*s* electrons. As illustrated in Fig. 2.2, the spin-down electrons are more likely reflected due to the strong scattering, while the spin-up electrons more likely get transmitted due to weak scattering. In summary, when unpolarized electrons transport through a ferromagnet with its majority spin

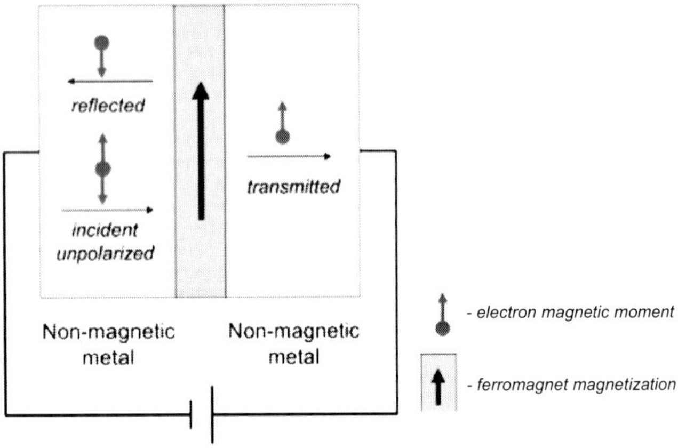

Figure 2.2 Spin-polarized electron transport through the nonmagnetic/ferromagnetic/nonmagnetic metal structure.

being up (down), the ferromagnet acts like a spin filter, favoring the spin-up (spin-down) current pass though, while blocking the spin-down (spin-up) electrons via stronger scattering. This serves as the foundation for STT in a subsequent section.

2.3 Giant and Tunneling Magnetoresistance

2.3.1 *Giant Magnetoresistance*

Major advances in electron spin transport started in the late 1970s and 1980s with the discovery of large low-temperature magnetoresistance in metallic superlattice [9, 10]. Later demonstrations of the "giant" effect at room temperature evolved toward applications to practical devices [11].

The first practical application of this phenomenon is in the giant magnetoresistance (GMR) effect. GMR is observed in thin-film materials composed of nanoscale alternate ferromagnetic and nonmagnetic layers (Fig. 2.3, left) [1], resulting in giant changes in the electrical resistance to an external magnetic field.

GMR was discovered in 1998 in the metallic thin-film multilayer structure with alternating ferromagnetic (Fe) and nonmagnetic (Cr)

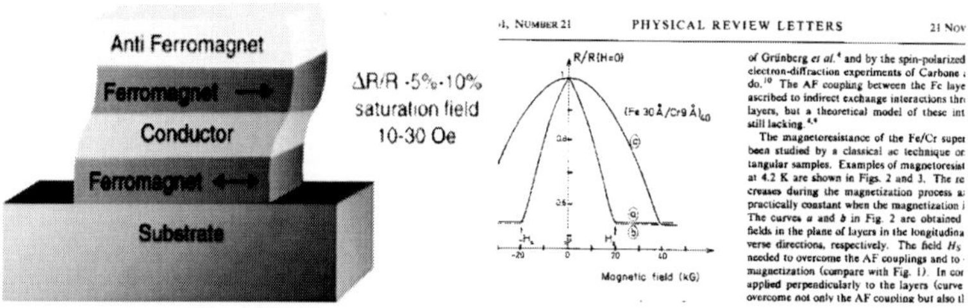

Figure 2.3 Structure of spin valve (left) [1] and magnetoresistance of three Fe/Cr superlattices at 4.2 K. Adapted from Ref. [10].

layers [10]. The discovery of the GMR effect won Drs. Albert Fert and Peter Grünberg the Nobel Prizes in Physics in 2007 [12].

The resistance (R_P) of the structure is relatively low when the magnetic orientations of those ferromagnetic layers are parallel, while the resistance (R_AP) becomes relatively high when those alternating ferromagnetic layers are antiparallel to each other. The magnetoresistance ratio (MR) is defined as the ratio of the change in resistance to the resistance in parallel configuration:

$$\mathrm{MR} = \frac{R_\mathrm{AP} - R_\mathrm{P}}{R_\mathrm{P}} \tag{2.2}$$

The resistance of the material is lowest when the magnetic moments in ferromagnetic layers are aligned in the same direction and highest when they are antialigned.

To understand this GMR phenomenon, we started with the simple three-layer structure that consists of two ferromagnetic layers sandwiched by a nonmagnetic metal layer. This ferromagnetic/nonmagnetic/ferromagnetic (F/N/F) structure is usually called a *spin valve*, that is, a device that gauges the movement of spins.

The GMR effect has been observed in both current-in-the-plane (CIP) and current-perpendicular-to-the-plane (CPP) configurations. Here we present the explanation of GMR using the CPP configuration. When electrons pass through the ferromagnet with the electron spin orientations parallel to its magnetization, $R_\downarrow$ when the spins of the electrons are antiparallel to the ferromagnet magnetization. When the magnetizations of the two ferromagnetic layers are parallel (P), the spin-up electrons have a combined resistance $2R_\uparrow$ and the spin-down electrons see a resistance of $2R_\downarrow$, so the total combined resistance in the parallel configuration is $R_\mathrm{P} = 2R_\uparrow R_\downarrow/(R_\uparrow + R_\downarrow)$. Similarly, in the antiparallel configuration of the two ferromagnets (AP), both spin-up and spin-down electrons bare the same resistance of $(R_\uparrow + R_\downarrow)$, so the overall resistance in the antiparallel configuration is $R_\mathrm{AP} = (R_\uparrow + R_\downarrow)/2$. When $R_\uparrow < R_\downarrow$, which is the case when you consider the spin-polarized transport discussed previously, we have R_AP higher than R_P, which qualitatively explains the phenomenon of GMR.

After GMR was discovered, it was immediately realized that this not only had the potential to enhance the performance of the hard disk drive (HDD). Spin valves operate at room temperature and exhibit significant changes in resistivity when subjected to relatively small external magnetic fields. The imposed magnetic field changes the magnetic orientation of one of the two layers, disrupting their relative orientation and thus changing the resistivity. Thus they can be used as magnetic field sensors as in the read head in the HDD, which entails a multibillion-dollar market annually. The market force accelerated the research as well as the miniaturization of GMR-based read heads. The first GMR-based magnetic field sensor was created in 1994, and high-performance disk drives utilizing GMR-based read heads to detect magnetic fields were realized in 1997 and now are ubiquitous, as illustrated in Fig. 2.4. These read heads are responsible for the very rapid growth in magnetic storage densities that has occurred in the last decade. The huge increase in the data storage density enables the significant decrease in cost per megabyte by the miniaturization of GMR read head and the improvements in magnetic materials. But for a very long period of time, the GMR read head technology was ahead of the development of the magnetic medium coatings, which has been catching up in recent years.

2.3.2 *Tunneling Magnetoresistance*

Spin-dependent tunneling was first discovered by Tedrow and Meservey in a series of experiments in ferromagnet/insulator/superconductor (F/I/S) junctions [13, 14]. These experiments

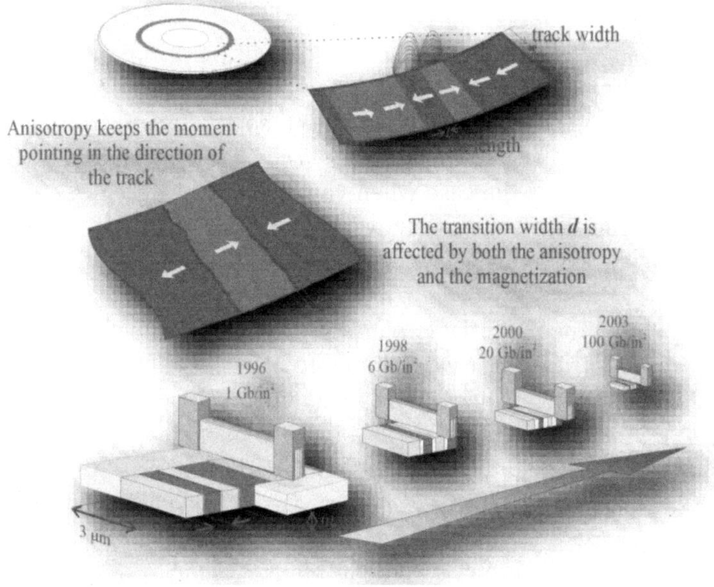

Figure 2.4 Schematic and evolution of GMR read heads since 1996. Adapted from Spintronics lecture notes, Prof. Stu Wolf, UVA 2006.

showed that conduction electrons in ferromagnetic materials are spin-polarized and the spin is conserved during the tunneling process.

In 1975, Julliere studied the tunneling conductance of ferromagnet/insulator/ferromagnet (F/I/F) where the insulator was an amorphous Ge layer. By applying the same analysis of the tunneling conductance of F/I/S from Tedrow and Meservey to F/I/F, he formulated a model to describe the change of conductance between the parallel and antiparallel magnetization of the two ferromagnetic layers [15]. The corresponding TMR in an F/I/F-type structure, called magnetic tunnel junction (MTJ), is defined as

$$\mathrm{TMR} = \frac{\Delta R}{R_\mathrm{P}} = \frac{R_\mathrm{AP} - R_\mathrm{P}}{R_\mathrm{P}} \tag{2.3}$$

where the resistances (R) are labeled by the relative orientation (P or AP) of the magnetizations of F1 and F2.

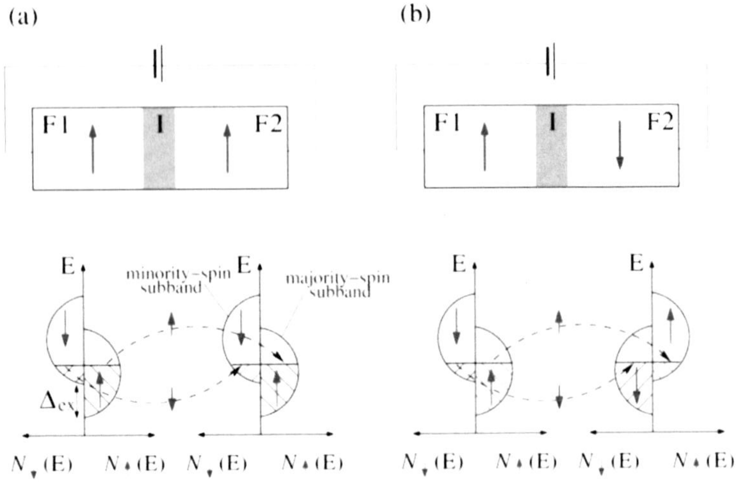

Figure 2.5 Schematics of electron tunneling in ferromagnet/insulator/ferromagnet (F/I/F) tunnel junctions. (a) Parallel and (b) antiparallel orientation of magnetizations with the corresponding spin-resolved density of the d states in ferromagnetic metals that have exchange spin splitting Δ_{ex}. Arrows in the two ferromagnetic regions are determined by the majority spin sub-band. Dashed lines depict spin-conserved tunneling. Reproduced from Ref. [8].

According to Ref. [16], the conductance (the reciprocal of the resistance R) could be expressed as $G_P = \frac{1}{R_P} = N_{M1}N_{M2} + N_{m1}N_{m2}$ and $G_{AP} = \frac{1}{R_{AP}} = N_{M1}N_{m2} + N_{m1}N_{M2}$, where N_{Mi} and N_{mi} are the spin-resolved density of states for majority and minority spin in the ferromagnets. Recalling the definition of spin polarization $P = \frac{N_\uparrow - N_\downarrow}{N_\uparrow + N_\downarrow}$ (Eq. 2.1), Eq. 2.3 could be rewritten as

$$\text{TMR} = \frac{2P_1P_2}{1 - P_1P_2} \tag{2.4}$$

2.4 Magnetic Tunnel Junction and MRAM

As introduced in the previous section, the trilayer junction of two ferromagnetic layers sandwiched by an insulating barrier is called an MTJ. The first experiment of spin-dependent tunneling in an MTJ-type structure Fe/Ge/Fe by Julliere in 1975 [15] was carried out at low temperature and a TMR effect of 14% was reported. This type of work was continued by Maekawa and Gafvert in 1982 [16] on a Ni/NiO/FM system, where FM stands for Fe, Co, or Ni. They were only able to observe a few percent of TMR in low temperature as well. The breakthrough came in 1995 when two groups, one from MIT [17] and the other from Japan [18], reported significant progress in making MTJs showing the room-temperature TMR effect. Moodera et al. achieved a TMR value of 24% at 4.2 K and 12% at room temperature in the structure of CoFe/Al_2O_3/Co (or NiFe) [17]. Similarly, Miyazaki and his collaborators obtained 30% TMR at 4.2 K and a room-temperature TMR of 18% from Fe/Al_2O_3/Fe MTJs [18].

The achievement of room-temperature TMR in 1995 attracted a great deal of interest from research groups across the world, mainly due to the huge potential industrial applications of TMR and the broader research field by exploring another degree of freedom of electrons, the spins of the electrons, besides the conventional manipulation of the charges of electrons. Hundreds of papers have been published on the subject of MTJs after the initial breakthrough in 1995 in order to improve the TMR performance of MTJs for industrial applications, among which the two major applications are using MTJs as HDD recording head sensors and nonvolatile magnetic random access memory (MRAM) cells, both of which will be discussed in a later section.

The initial breakthrough on MTJs with simple electrodes like CoFe and Fe produced a TMR just over 10%. Material-wise, by replacing polycrystalline electrodes like CoFe and NiCoFe with amorphous CoFeB and optimizing the growth conditions, a record-high magnetoresistance value of 70% was achieved on MTJs with an aluminum oxide barrier [19]. However, this was still not enough for many spintronic device applications. The breakthrough came when the first-principle calculations predicted that MTJs with an epitaxial Fe/MgO/Fe structure could provide significant spin filtering that would enhance the magnetoresistance to over 1000% [20], and was soon demonstrated experimentally with a TMR around 200% by two different groups independently [21, 22]. Combing the CoFeB electrode and a MgO tunnel barrier, a magnetoresistance value of 230% at room temperature was achieved in MTJs with a CoFeB/MgO/CoFeB structure [23]. By optimizing the composition

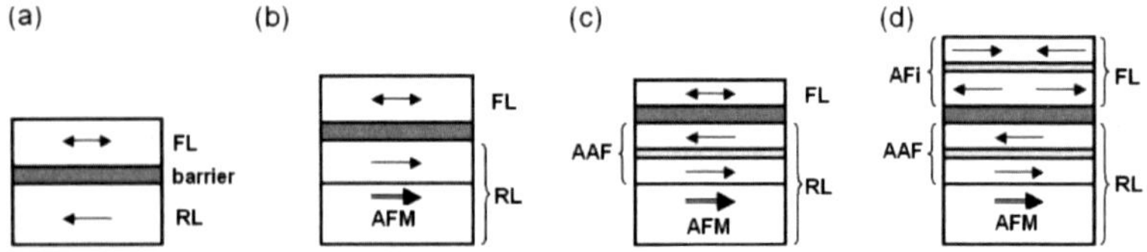

Figure 2.6 The evolution of MTJ structures. (a) MJTs with a single ferromagnetic free layer (FL), whose magnetization could be rotated to achieve high and low resistance states, and a reference layer (RL), whose magnetization is fixed. (b) MTJs with an exchange-biased RL. (c) Exchange-biased synthetic antiferromagnet (SAF) as the RL. (d) Exchange-biased SAF as the RL and an artificial ferromagnet (AFi) as the FL of MTJs. Adapted from Ref. [25].

of CoFeB, recently TMR ratios of up to 500% have been achieved at room temperature in CoFeB/MgO/CoFeB MTJs [24]. All of these efforts have greatly enhanced the potential of MTJs in spintronic device applications like data storage, magnetic sensors, etc.

Besides the improvement of materials for making better MTJs, different layered structures have also been intensively studied to produce MTJs more suitable for practical applications. As shown in Fig. 2.6, the initial MTJs have the simplest structure of two ferromagnetic layers sandwiched by an ultrathin insulating barrier layer. One of the ferromagnetic layers has relatively large magnetic anisotropy and its magnetization direction is fixed as the reference orientation, so this ferromagnetic layer is called the reference layer (RL). The other ferromagnetic layer, called the free layer (FL), has relatively low coercivity and hence its magnetization could be flipped in one of the two directions easily in order to achieve the relative high and low resistance of the whole MTJ stack.

To achieve a more stable RL, the exchange bias (EB) mechanism was utilized by adding an antiferromagnetic (AFM) layer next to the ferromagnetic RL to form the bilayer EB system. Basically, due to the EB coupling between the ferromagnetic/AFM layers, the hysteresis loop of the EB system will be shift off center along the field direction caused by the EB anisotropy. The details of the EB system can be found in Refs. [26]. Here by shifting the hysteresis loop of the RL, a more stable and wider range separation of the two ferromagnetic layers was achieved. To reduce the magnetic stray field from the RL and further stabilize the RL, the ferromagnetic layer in the EB system was replace by a trilayer system with two ferromagnetic layers sandwiched by a nonmagnetic metal layer through the AFM coupling, as shown in Fig. 2.6c. This sandwiched system is called a synthetic antiferromagnet (SAF) or an artificial ferrimagnet (AFi), depending on its net magnetic moment [25]. Later, such a sandwiched system by AFM coupling was also used as the FL of MTJs, with the main purpose of reducing the stray field [27].

MRAM is based on the concept of using the direction of magnetization to store information and magnetoresistance (variation in resistance with magnetization) as the information readout. It was originally invented in the late 1980s to replace bulky and heavy plated wire radiation hard memories. MRAM was a truly nonvolatile (i.e., retains memory when power is off) memory device for specialty radiation hard applications when it was first invented, with dimensions larger than 1 micron.

Higher TMR in MTJ materials [17, 18] enabled, for the first time, one MTJ element and one transistor bit cell (1T-MTJ) architecture, as shown in Fig. 2.7. The magnetoresistance of an MTJ (TMR ratio) is much higher than for GMR material, and the read signal of MTJ-based MRAM is greatly improved as a result. A read speed in the range of 10 ns became feasible as a consequence of the improved signal levels.

The writing of standard MRAM is accomplished by flowing orthogonal currents to produce the required orthogonal magnetic fields at the cell at their cross point. The writing current required per line is on the order of 5–10 mA or larger: a very large pass transistor is required to control the pulsing of this large current. The only feasible option to minimize memory chip footprint is to share this large write current over many rows or columns of cells. The bits at the cross points get written, but the bits that experience half of the writing field under just one current line are "half selected." The magnetic switching field of a single free-layer bit depends sensitively on the exact bit shape, since shape anisotropy is the main contributor to the free-layer magnetic anisotropy, and has a large variation in a large memory array. As a result, some bits may experience half-select disturb switching problems. This was one

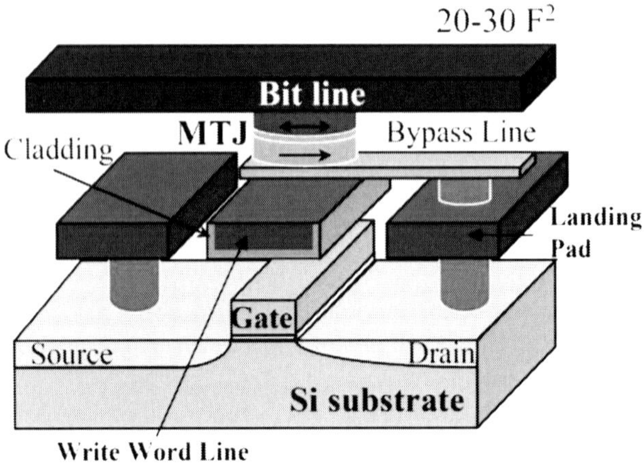

Figure 2.7 Schematics of 1T-1MTJ MRAM. Adapted from Ref. [29].

of the key problems blocking the development of MRAM memory products. A new storage mode using an SAF FL was invented and resulted in a rotational switching of the FL with a greatly enhanced writing operational window [28].

An MRAM chip product prototype was announced in 2003 and the first-ever MRAM commercial products started to ship in 2006 [30]. The manufacturability of MTJ devices as memory bits was proven in these nonvolatile MRAM products, having high read and write speeds (tens of nanoseconds) and nearly unlimited endurance.

Today 180 nm and 90 nm stand-alone MRAM products are finding more and more applications in various areas where their unique features of being nonvolatile fast and of unlimited endurances are vital: satellite applications, automotive data recorders, industrial controls, etc. Efforts are ongoing to scale MRAM technology to 65 nm for both stand-alone and embedded applications.

2.5 Spin Torque Transfer

In 1996, Slonczewski and Berger predicted that spin-polarized current could exert a spin torque on the magnetic moments when passing through a ferromagnetic layer, generating spin waves, or even completely switch the magnetization orientation

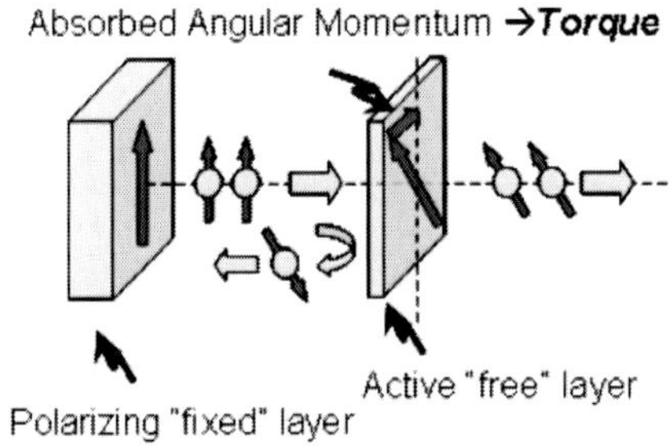

Figure 2.8 Spin torque transfer (STT) inside a spin valve or magnetic tunnel junction scheme. When an electron current passes through the ferromagnetic fixed layer, it becomes spin polarized, and when flowing through the second ferromagnetic free layer, it exerts a spin torque on the magnetic moment of the free layer due to the interaction between the electron spin and local magnetization. Adapted from Spintronics lecture notes, Prof. Stu Wolf, UVA 2006.

of the ferromagnetic layer, if properly engineered [31, 32]. This prediction was based on a generally accepted argument: For a system consisting of itinerant electrons and local moments, the total angular momentum is conserved, even when the system is out of equilibrium. It was predicted by Slonczewski that the spin torque from a direct current (DC) could induce two different types of magnetic behaviors, switching from one magnetic orientation to another and a persistent precession of magnetization.

The STT induced the magnetic switching becomes the underpinning technology that enabled the next generation of MRAM, thanks to the scalability and the simple architecture. The STT switching mechanism is illustrated in Fig. 2.8 within a spin valve or MTJ structure frame.

Compared to the use of the external field generated by passing current in a conventional toggle switch scheme, the STT switching scheme has the correct scaling behavior; namely, it is the current density rather than the absolute current that is important, so the actual current scales down properly with the device size. The theoretical expression of this critical current density required to switch the magnetization of a ferromagnet is predicted by Slonczewski [31]:

$$J_C = \frac{2e\alpha \mathbf{M}_S t(\mathbf{H}_K \pm \mathbf{H} + 2\pi \mathbf{M}_S)}{\hbar \eta} \quad (2.5)$$

where e is the electron charge, α is the damping constant, $\mathbf{M}_S$ is the saturation magnetization, t_F is the thickness of the FL, $\mathbf{H}_K$ is the effective anisotropy field including magnetocrystalline anisotropy and shape anisotropy, $\mathbf{H}_{ext}$ is the external field, $\hbar$ is the reduced Planck's constant, and η is the spin transfer efficiency. The $\pm$ sign denotes the positive and negative directions of the electron current flow, with the positive direction defined as the electron flow from the fixed RL to the FL.

The STT switching effect was initially demonstrated in the metallic spin valve nanopillars, showing MR values between 0.5% and 5% and the critical current density J_C ranging from 5×10^6 to 5×10^8 A/cm^2 in various structures [33–35]. Later the focus was shifted to the study of STT switching of MTJs, which is more closed to practical applications due to their much larger TMR values, as a high TMR value is needed for fast read operation in the MRAM application. The first experimental demonstration of STT switching of an MTJ was achieved by a group from Grandis Inc., which demonstrated the effect in MTJs with 42% TMR and a resistance–area (RA) product value $\sim 40\ \Omega \cdot \mu m^2$. The critical current density they obtained was around 6×10^6 A/cm^2 [36].

2.6 Spin Torque Transfer Applications

2.6.1 *STT-RAM*

The STT switching technique brings significant advantages to the next-generation MRAM called STT-RAM in order to be distinguished from the conventional MRAM. In STT-RAM, the half-select disturb problem of conventional field writing MRAM was avoided because the writing currents only flow to the cells being written. STT-RAM enables MRAM scalability beyond 65 nm by reducing the write current more than 10× in a 1T (transistor)/1MTJ memory cell architecture such as the one shown in Fig. 2.9. It also leads to simpler memory architecture and manufacturing than conventional MRAM.

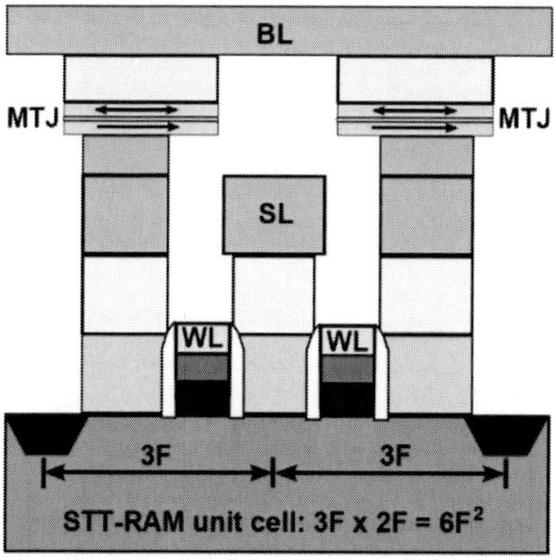

Figure 2.9 Schematics of STT-RAM. Adapted from Ref. [29].

A viable new memory technology must demonstrate a clear path to migrate to smaller and denser memory size with lower power consumption as the underlying complementary metal–oxide semiconductor (CMOS) logic technology scales down, as predicted by the International Technology Roadmap for Semiconductors (ITRS) roadmap. One of the major drawbacks of conventional MRAM has been the increase of switching current as the technology scales down. STT-RAM solves this switching current problem by applying the spin-polarized current vertically through the MTJ element. Whether the nanomagnet switches or not is determined roughly by the current density; therefore, as the CMOS technology scales down, and the area of the MTJ element decreases, the total current decreases.

One of the key issues of STT-RAM technology has always been the reduction of the STT writing current of the MTJ storage element while maintaining sufficient thermal stability for adequate data retention and minimal write/read error rates [3]. Thermal stability of the storage layer in an MTJ not only affects STT-RAM memory cell stand-by data retention, it also affects the write current distribution and hence the write error rate and the read disturb error rate.

To switch the FL magnetization in a conventional in-plane MTJ device, a spin-polarized current is applied, and the FL magnetization starts precessing around the direction of the total effective field. The magnetization has to overcome a very large out-of-plane demagnetizing field before it can switch to the opposite direction. The out-of-plane demagnetizing field does not contribute to the thermal stability of the in-plane FL, yet it causes the STT writing current to be much higher, thus making the ratio of the writing current over Δ higher. Perpendicular magnetic tunnel junctions (PMTJs) for STT-RAM had been investigated to solve this problem [37, 38]. Indeed a very low STT writing current was demonstrated at a reasonably high thermal stability factor [39]. Perpendicular magnetic materials have been extensively studied for HDD application as a recording media. This type of material is capable of having a very high magnetic anisotropy field arising from the atomic properties of these materials. No shape anisotropy is required, and circular-shaped bits can be used, which simplifies the manufacturing process. Sizes below even 10 nm are expected to have enough thermal stability due to the high anisotropy field (>2 Tesla) in some of these materials. The main challenges of making PMTJ work for STT-RAM are how to make these materials compatible with the MgO tunnel barrier, yielding a high TMR ratio and spin transfer efficiency and decreasing their damping constants for lower STT writing currents.

Another promising approach in solving the high out-of-plane demagnetizing field of conventional in-plane MTJ devices is to induce strong perpendicular anisotropy in the in-plane free-layer material through material engineering, to the degree that it nearly cancels the demagnetizing field so that the in-plane magnetization of the FL can easily rotate out of plane and switch to the opposite in-plane direction [40]. Shape anisotropy will still be needed to achieve adequate thermal stability.

Today the lowest STT critical writing current density reported with reasonable storage thermal stability (>40) was about 1 to 2×10^6 A/cm^2 for in-plane MTJ materials [40] and around 2×10^6 A/cm^2 for PMTJ materials [41]. Because in-plane MTJ elements need an aspect ratio of more than 2 to 3, and the PMTJ simply uses a circular shape, total STT writing currents are comparable. As shown in Fig. 2.9, such write current levels enable the design of smaller than 8 F^2 STT-RAM memory cells.

For this current drive STT switching to be viable in the MRAM architecture, the critical current density needs to be reduced to the level of $\sim 10^5$ A/cm^2 for a few reasons. First of all, the switching current in the MRAM architecture is delivered by a CMOS transistor, which typically supplies 100 μA per 100 nm gate width. To achieve higher-density MRAM with 1T-1MTJ architecture, the size of the transistor needs to be reduced, resulting in smaller current output from the transistor and hence a smaller switching current density becomes inevitable. Also, a large current density will produce a large voltage potential across the device, increasing the probability of material fatigue and tunnel barrier breakdown, compromising the endurance and reliability of the overall device performance [32].

At the present time, a high TMR ratio $>150\%$ and low RA ($<10\ \Omega \cdot \mu m^2$) seem to be achieved by various groups using in-plane MTJ materials. In the case of PMTJ materials, on the other hand, it seems hard to achieve these transport performances. More work is needed in this area.

Future STT-RAM architectures employing cross-point architecture and multiple (>4) storage states per bit can effectively lower the memory cell size to the 2–4 F^2 range to compete with storage class memories such as Flash.

The switching of an MTJ FL using STT is intrinsically capable of being completed in less than 1 ns if the writing current amplitude is large enough. A typical writing current dependence on writing current pulse width is shown below in Fig. 2.10 for an in-plane MTJ elliptical element with a size of 90 $\times$ 180 nm. It can be seen that there is a trade-off between the writing speed and the current level required, which scales with the size of the transistor or memory cell size.

The read operation of STT-RAM is very similar to conventional field writing MRAM; even an MTJ based on an AlO_x tunnel barrier, with a much lower TMR ratio ($\sim 25\%$), supports a read access time of 30 ns. STT-RAM uses a MgO barrier–based MTJ having a TMR ratio greater than 120%. Circuit simulation showed that a 2–5 ns read access time is possible with even a 2–5 μA read current differential. Unlike MRAM, STT-RAM has a read current disturb switching issue. This can be minimized by controlling the read current level and designing the bit with proper thermal stability [42].

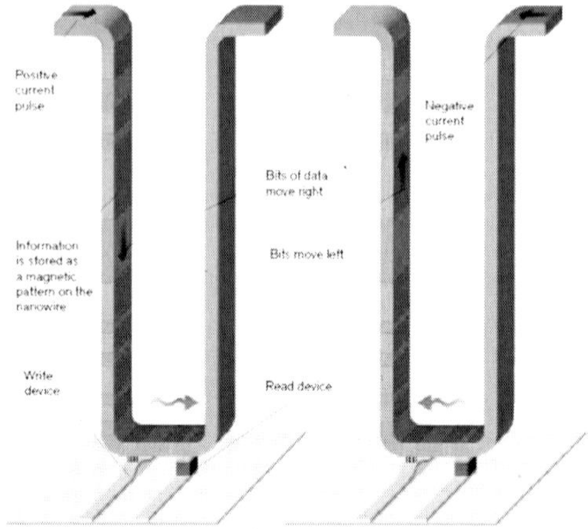

Figure 2.10 Schematic of domain wall racetrack memory. Adapted from Ref. [45].

STT-RAM retains the major benefits of MRAM of being a fast and nonvolatile memory with no known wear-out mechanism, and unlimited endurance, and extends the scalability to sub-10 nm nodes [29]. STT-RAM is being demonstrated at a leading-edge CMOS node like other mainstream semiconductor memories.

2.6.2 *Current-Induced Domain Wall Motion Memory*

The concept of storing information in movable magnetic domain walls (MDWs) was conceived more than three decades ago, which triggered the intensive research in *magnetic bubble memories* [43]. The magnetic bubbles were often arranged in the form of shift registers and magnetic fields were required to operate. However, this required on-chip field generators at the same size scale as the individual magnetic bits, adding considerable complexity and cost to the device and making scaling to smaller dimensions very difficult [44].

Current-driven magnetic domain wall motion (MDWM), first predicted by Berger [32], has been intensively studied both experimentally and theoretically in recent years. Applications of current-driven MDWs to storage logic and memory devices have been also presented. Among them, the racetrack memory proposed by Stuart Parkin at IBM received considerable attention [45], as illustrated in Fig. 2.10. In Permalloy (NiFe) nanostrips, in which critical current density has been reported to be around $\sim 10^8$ A/cm^2 and the re-domain wall velocity was $\sim$150 m/s [46]. The current density is still 2 orders of magnitude too large for practical applications. However, a relatively small critical current density has been reported for nanostrips with perpendicular magnetic anisotropy (PMA) [41]. In addition, perpendicular magnetic thin films exhibited much narrower one-dimensional MDWs that can strongly interact with defects.

Recently, current-driven MDWM was also demonstrated in conjunction with MTJs for MRAM. Similar to STT-RAM, MDWM devices require a spin-polarized current to control the positions of MDWs and this has the desirable scalability, very low power, and nonvolatility.

Wolf et al. proposed a flux shuttle MRAM that combines multiferroic materials with MDW devices [47], as shown in Fig. 2.11. A multiferroic layer, for example, $BiFeO_3$, acts as an EB layer that pins the FL at intervals. If a gate voltage is applied to any multiferroic segment then the EB is reduced and the FL is not pinned as strongly (this principle has been demonstrated by Chu et al. [40]) so that a small spin torque current is sufficient to rotate the magnetization direction of the FL by moving the domain wall from one notch to the one on the other side of the junction. To write the individual bit, a spin-polarized current is passed through the common magnetic layer (the blue layer in the Fig. 2.11) and the domain wall can only move with a voltage applied to the multiferroic adajent to the common magnetic layer. The direction of the domain wall motion is dependent to the current flow direction. Thus many bits in the common magnetic layer can be written simultaneously depending on the selection of bits in the series array of junctions. A key feature of this memory is the ability to store multiple states in each bit. Since the resistance of the MTJ depends on the exact position of the domain wall in the FL or in the layer in close proximity to the FL. This position can be controlled by the timing and duration of the voltage pulse on the multiferroic relative to the current pulse in the FL or storage layer; thus the resistance states can be discrete or continuous.

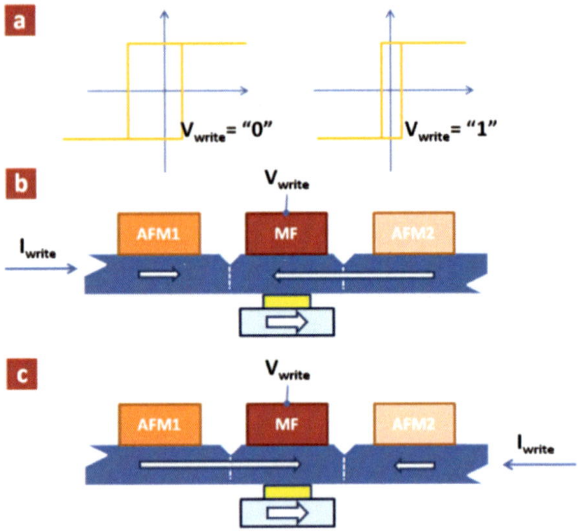

Figure 2.11 (a) Hysteresis loops of the free layer underneath the multiferroic ($BiFeO_3$) layers. The EB is formed due to the coupling of the ferromagnetism of the free layer and the antiferromagnetism of BFO [40]. The application of a gate voltage onto the MF lifts the EB so that the ferromagnetism in the free layer is more susceptible to change polarity. (b) Write "1": the spin-polarized current travels from left to right and the spin polarization is determined by the magnetic domain to the left of the MTJ. (c) Write "0": the current travels from right to left [47].

2.6.3 *Spin Torque Oscillator*

As mentioned earlier, the spin torque from a spin-polarized current could excite a dynamical precession state of the magnetization. The devices that undergo precession are called spin torque oscillators (STOs), as shown in Fig. 2.12. For sufficiently dense spin currents, this spin torque can become large enough to overcome the intrinsic magnetic damping effect in the FL, and drive magnetization motion. In an STO, this spin transfer effect is used to drive the magnetization of a FL that *precesses* in a coherent fashion. A DC electron current produces a spin torque that counteracts the damping torque, and results in the magnetization precessing around the net effective field (Fig. 2.11). The GMR in a spin valve, or the TMR in an MTJ, will then lead to an alternating current (AC) signal superimposed over the

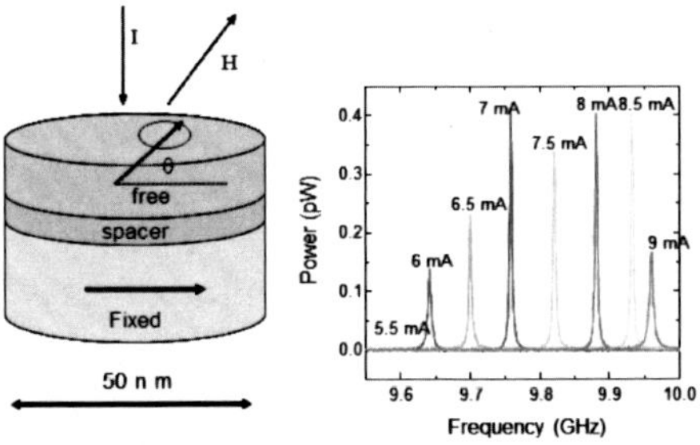

Figure 2.12 Schematic of a spin torque oscillator (STO) (left) and the frequency of oscillation of the STO for various injection currents (right). Adapted from Ref. [48].

applied DC current with a frequency determined by the frequency of precession and that can be detected electrically, as demonstrate by Ralph and his collaborators [48].

In 1998 Tsoi et al. [49] demonstrated indirect evidence of microwave emission in magnetic point contact structures in which a DC current is injected through a narrow nanoscale (tens of nanometers) contact on top of the unpatterned FL. Direct observation of microwave output from spin transfer devices was achieved at Cornell [48] and the National Institute of Standards and Technology (NIST) [50]. The NIST group, using a lithographed point contact geometry, demonstrated narrow linewidths ($<$2 MHz), power outputs close to the expected values, and a high degree of current and field tunability. These devices have been shown to exhibit high frequency (0.1–50 GHz)-coherent oscillations. Several groups [51, 52] subsequently demonstrated mutual phase locking of two closely spaced STOs and demonstrated a fourfold (the square of the number of oscillators) increase in power when the STOs are phase-locked.

The initial STO research used spin valve structures based on the GMR effect with very low impedances and with GMR ratios of only a few percent. At typical operating currents of $I_{\text{dc}} = 10$ mA and load resistances $R_{\text{L}} = 50$ these values limited the output power to 2 nW. STOs with considerably larger power and efficiencies have become feasible due to recent progress in the development of MgO-based MTJs that show large TMRs. The use of MTJs also allows for the

STO impedance to be increased to 50 to 5000 Ω. This large tunable impedance is advantageous when trying to match to an incoming signal or to an antenna. STO device technologies using MTJs are also very attractive due to the similarity of STO devices to HDD read heads and MRAM. This allows STO technologies to leverage both the innovations and the mature fabrication processes developed for these large related industries.

There is an incredible range of possible applications of STOs. The nanoscale dimensions, coupled with the possibility of CMOS integration, low power, tunable frequency through magnetic fields and electrical currents, high quality factor Q (>10,000), and the control of coupling using multiferroics, opens many possibilities, from RF applications, filters, and mixers to on-chip clock generation. As nanoscale microwave oscillators, STOs offer many features that make them attractive for microwave applications. First STOs are highly tunable: the output frequencies are a function of the applied field and current. By injecting an incoming signal into the STO current (or local field) the incoming signal can be mixed with the reference oscillator signal. The STO therefore functions as a nanoscale local oscillator and mixer. The nanoscale nature of these devices allows large numbers to be fabricated in small areas. Further, the nanoscale nature enables the devices to be located near the incoming signal, such as directly integrated onto a coplanar wave guide or microstrip line, without perturbing the signal with capacitive loading.

References

1. S. A. Wolf, Spintronics: a spin-based electronics vision for the future, *Science*, **294**, 1488–1495 (2001).
2. D. E. Nikonov, G. I. Bourianoff, and P. A. Gargini, Power dissipation in spintronic devices out of thermodynamic equilibrium, *J. Supercond.*, **19**, 497–513 (2006).
3. E. Chen et al., Advances and future prospects of spin-transfer torque random access memory, *IEEE Trans. Magn.*, **46**, 1873–1878 (2010).
4. X. W. Wu, Nonmagnetic shell in surfactant-coated FePt nanoparticles, *J. Appl. Phys.*, **95**, 6810 (2004).

5. J. K. Kübler, *Theory of Itinerant Electron Magnetism*, Clarendon Press, 2000.
6. E. C. Stoner, Atomic moments in ferromagnetic metals and alloys with non-ferromagnetic elements, *Philos. Mag. Series 7*, **15**, 1018–1034 (2012).
7. N. F. Mott, The electrical conductivity of transition metals, *Proc. R. Soc. A: Math., Phys. Eng. Sci.*, **153**, 699–717 (1936).
8. I. Zutic, J. Fabian, and S. Das, Spintronics: fundamentals and applications, *Rev. Mod. Phys.*, **76**, 323–410 (2004).
9. I. Schuller et al., Transport properties of the compositionally modulation alloy Cu/Ni, *AIP Conf. Proc.*, **53**, 417–421 (1979).
10. M. Baibich et al., Giant magnetoresistance of(001) Fe/(001) Cr magnetic superlattices, *Phys. Rev. Lett.*, **61**, 2472–2475 (1988).
11. G. A. Prinz, Magnetoelectronics, *Science*, **282**, 1660–1663 (1998).
12. E. E. Fullerton and I. K. Schuller, The 2007 Nobel Prize in physics: magnetism and transport at the nanoscale, *ACS Nano*, **1**, 384–389 (2007).
13. P. Tedrow and R. Meservey, Spin-dependent tunneling into ferromagnetic nickel, *Phys. Rev. Lett.*, **26**, 192–& (1971).
14. P. Tedrow and R. Meservey, Spin polarization of electrons tunneling from films of Fe, Co, Ni, and Gd, *Phys. Rev. B*, **7**, 318 (1973).
15. M. Julliere, Tunneling between ferromagnetic films, *Phys. Lett. A*, **54**, 225–226 (1975).
16. S. Maekawa and U. Gafvert, Electron-tunneling between ferromagnetic-films, *IEEE Trans. Magn.*, **18**, 707–708 (1982).
17. J. Moodera, L. Kinder, T. Wong, and R. Meservey, Large magnetoresistance at room-temperature in ferromagnetic thin-film tunnel-junctions, *Phys. Rev. Lett.*, **74**, 3273–3276 (1995).
18. T. Miyazaki and N. Tezuka, Giant magnetic tunneling effect in Fe/Al2O3/Fe junction, *J. Magn. Magn. Mater.*, **139**, L231–L234 (1995).
19. D. Wang et al., 70% TMR at room temperature for SDT sandwich junctions with CoFeB as free and reference layers, *IEEE Trans. Magn.*, **40**, 2269–2271 (2004).
20. W. Butler, X. G. Zhang, T. Schulthess, and J. MacLaren, Spin-dependent tunneling conductance of Fe|MgO|Fe sandwiches, *Phys. Rev. B*, **63**, (2001).
21. S. S. P. Parkin et al., Giant tunnelling magnetoresistance at room temperature with MgO (100) tunnel barriers, *Nat. Mater.*, **3**, 862–867 (2004).

22. S. Yuasa, T. Nagahama, A. Fukushima, Y. Suzuki, and K. Ando, Giant room-temperature magnetoresistance in single-crystal Fe/MgO/Fe magnetic tunnel junctions, *Nat. Mater.*, **3**, 868–871 (2004).
23. D. D. Djayaprawira et al., 230% room-temperature magnetoresistance in CoFeB/MgO/CoFeB magnetic tunnel junctions, *Appl. Phys. Lett.*, **86**, 092502 (2005).
24. Y. M. Lee, J. Hayakawa, S. Ikeda, F. Matsukura, and H. Ohno, Effect of electrode composition on the tunnel magnetoresistance of pseudo-spin-valve magnetic tunnel junction with a MgO tunnel barrier, *Appl. Phys. Lett.*, **90**, 212507 (2007).
25. N. Wiese, Coupling phenomena and scalability of CoFeB/Ru/CoFeB sandwiches, Bielefeld (Germany): Bielefeld University (2006).
26. J. Nogues and I. K. Schuller, Exchange bias, *J. Magn. Magn. Mater.*, **192**, 203–232 (1999).
27. S. Yuasa, Giant tunneling magnetoresistance in MgO-based magnetic tunnel junctions, *J. Phys. Soc. Jpn.*, **77**, 031001 (2008).
28. T. Maffitt et al., Design considerations for MRAM, *IBM J. Res. Dev.*, **50**, 25–39 (2005).
29. S. A. Wolf, J. Lu, M. R. Stan, E. Chen, and D. M. Treger, The promise of nanomagnetics and spintronics for future logic and universal memory, *Proc. IEEE*, **98**, 2155–2168 (2010).
30. B. N. Engel et al., A 4-Mb toggle MRAM based on a novel bit and switching method, *IEEE Trans. Magn.*, **41**, 132–136.
31. J. Slonczewski, Current-driven excitation of magnetic multilayers, *J. Magn. Magn. Mater.*, **159**, L1–L7 (1996).
32. L. Berger, Emission of spin waves by a magnetic multilayer traversed by a current, *Phys. Rev. B*, **54**, 9353–9358 (1996).
33. F. J. Albert, J. A. Katine, R. A. Buhrman, and D. C. Ralph, Spin-polarized current switching of a Co thin film nanomagnet, *Appl. Phys. Lett.*, **77**, 3809 (2000).
34. J. A. Katine, F. J. Albert, R. A. Buhrman, E. B. Myers, and D. C. Ralph, Current-driven magnetization reversal and spin wave excitations in Co/Cu/Co pillars, *Phys. Rev. Lett.*, **84**, 3149 (2000).
35. J. Z. Sun, D. J. Monsma, D. W. Abraham, M. J. Rooks, and R. H. Koch, Batch-fabricated spin-injection magnetic switches, *Appl. Phys. Lett.*, **81**, 2202 (2002).
36. Y. Huai, F. Albert, P. Nguyen, M. Pakala, and T. Valet, Observation of spin-transfer switching in deep submicron-sized and low-resistance magnetic tunnel junctions, *Appl. Phys. Lett.*, **84**, 3118 (2004).

37. H. Meng and J.-P. Wang, Spin transfer in nanomagnetic devices with perpendicular anisotropy, *Appl. Phys. Lett.*, **88**, 172506 (2006).
38. X. Zhu and J. G. Zhu, Spin torque and field-driven perpendicular MRAM designs scalable to multi-Gb/chip capacity, *IEEE Trans. Magn.*, **42**, 2739–2741.
39. T. Kishi et al., Lower-current and fast switching of a perpendicular TMR for high speed and high density spin-transfer-torque MRAM, *Electron Devices Meeting, 2008. IEDM 2008. IEEE Int.*, 1–4 (2008).
40. Y.-H. Chu et al., Electric-field control of local ferromagnetism using a magnetoelectric multiferroic, *Nat. Mater.*, **7**, 478–482 (2008).
41. D. Ravelosona, S. Mangin, J. A. Katine, E. E. Fullerton, and B. D. Terris, Threshold currents to move domain walls in films with perpendicular anisotropy, *Appl. Phys. Lett.*, **90**, 072508 (2007).
42. E. Chen et al., Advances and future prospects of spin-transfer torque random access memory, *IEEE Trans. Magn.*, **46**, 1873–1878.
43. A. P. Malozemoff and J. C. Slonczewski, *Magnetic Domain Walls in Bubble Materials*, New York: Academic Press, 1979.
44. S. Middelhoek and P. George, *Physics of Computer Memory Devices* (1976).
45. S. S. P. Parkin, M. Hayashi, and L. Thomas, Magnetic domain-wall racetrack memory, *Science*, **320**, 190–194 (2008).
46. M. Hayashi, L. Thomas, R. Moriya, C. Rettner, and S. S. P. Parkin, Current-controlled magnetic domain-wall nanowire shift register, *Science*, **320**, 209–211 (2008).
47. S. Wolf and J. Lu, *Magnetic Random Access Memory Device*, WO Patent WO/2011/143,356 (2011).
48. S. Kiselev et al., Microwave oscillations of a nanomagnet driven by a spin-polarized current, *Nature*, **425**, 380–383 (2003).
49. M. Tsoi et al., Excitation of a magnetic multilayer by an electric current, *Phys. Rev. Lett.*, **80**, 4281–4284 (1998).
50. W. Rippard, M. Pufall, S. Kaka, S. Russek, and T. Silva, Direct-current induced dynamics in Co90Fe10/Ni80Fe20 point contacts, *Phys. Rev. Lett.*, **92**, (2004).
51. S. Kaka et al., Mutual phase-locking of microwave spin torque nano-oscillators, *Nature*, **437**, 389–392 (2005).
52. F. B. Mancoff, N. D. Rizzo, B. N. Engel, and S. Tehrani, Phase-locking in double-point-contact spin-transfer devices, *Nature*, **437**, 393–395 (2005).

Chapter 3

$L1_0$ Alloy Thin Films and Nanostructures

Jonathan R. Skuza
Center for Materials Research, Norfolk State University, 700 Park Avenue, Norfolk, VA 23504, USA
jrskuza@nsu.edu

Recently, a number of ferromagnetic (e.g., FePt, FePd, FeAu, CoPt, CoPd, NiPt, MnAl) and antiferromagnetic (e.g., MnNi, MnPd, MnPt, CrPt) thin-film and nanostructured materials that form the ordered $L1_0$ crystal structure have garnered significant attention, with FePt, FePd, and CoPt alloys leading the forefront of the research. Studies aimed at understanding the fundamental physics and potential uses of these $L1_0$ ordered materials in thin-film and nanostructured forms for various technological applications such as data storage (e.g., ultrahigh-density magnetic hard disk drives) have been ever increasing in the research literature. These $L1_0$ ordered alloys boast a very large magnetic anisotropy, large saturation magnetization, large coercivity, and excellent chemical stability. They also exhibit a magnetic easy axis oriented along the c axis (001) direction, which is advantageous for perpendicular magnetic media. Furthermore, these properties can be exploited and finely tuned through various deposition and synthesis techniques. To better understand these materials, various aspects of $L1_0$ ordered alloys in thin-film and

Handbook of Nanomagnetism: Applications and Tools
Edited by Rosa A. Lukaszew

ISBN 978-981-4613-04-0 (Hardcover), 978-981-4613-05-7 (eBook)
www.panstanford.com

nanostructured forms will be discussed in a logical way throughout this chapter. Beginning with the crystal structure of the disordered and ordered phases, an understanding of $L1_0$ chemical ordering from a physical and mathematical viewpoint will be emphasized. Naturally, this takes the discussion from the $L1_0$ structure to the magnetic properties of these alloys, since the two are strongly correlated. From here, various deposition and synthesis techniques that can create high-quality thin films and nanostructures will be introduced and the discussion will conclude with the use of these and other novel $L1_0$ ordered materials for magnetic data storage applications.

3.1 Introduction

3.1.1 *Data Storage*

The ability to record information and retrieve it at will "forever" or at least for a very long time has been important to the human race so that knowledge could be preserved and shared with future generations. This has become even more important in the digital age, where the volume of data that is created ($\sim$1.8 zettabytes in 2011) and the speed at which it needs to be retrieved have increased exponentially.

This is where magnetic data storage, in particular magnetic hard disk drives (HDDs), has come to the rescue because it can hold an immense volume of data for a very long time. Since IBM introduced the first HDD in 1956, the storage (or areal) density of magnetic HDDs has increased exponentially, as depicted in Fig. 3.1 [1] due to various technological breakthroughs, such as continuous-thin-film media, giant magnetoresistance (GMR), and antiferromagnetically coupled (AFC) media, as well as improvements in other recording aspects (e.g., signal processing, downscaling of components).

Now all of these technological breakthroughs through the early 2000s dealt with longitudinal magnetic recording (LMR), where the magnetic bits that store the 1s and 0s lie in the plane of the magnetic recording media. The size of these magnetic bits (i.e., grain size of the magnetic media) physically determines the fundamental

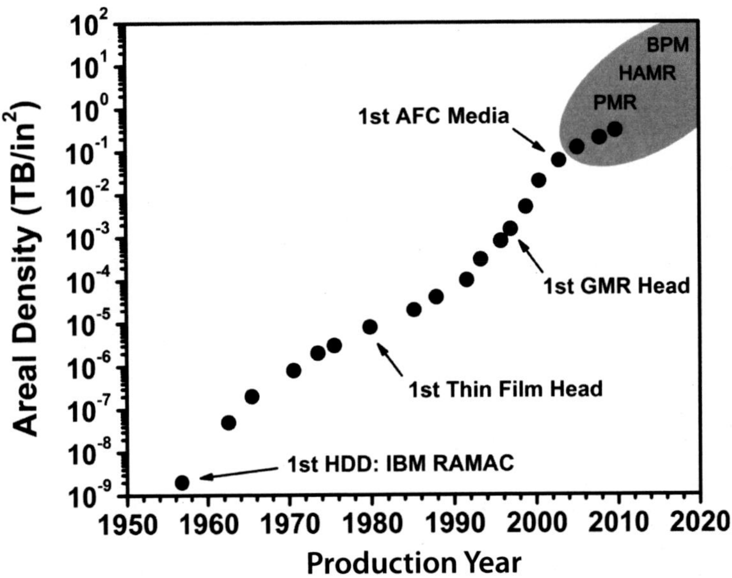

Figure 3.1 A perspective of the exponential increase of areal density in magnetic hard disk drives since their inception [1].

limit of the areal density (i.e., superparamagnetism), which LMR has practically reached by downscaling system components. Therefore, a new method of storing data on magnetic HDDs was needed and perpendicular magnetic recording (PMR) emerged in commercial devices [2], with magnetic bits lying out of the plane of the recording media. PMR has further allowed an increase of the areal density mainly due to the development of new materials for the recording media [2, 3]. Thus, PMR has temporarily allowed the postponement of the magnetic recording trilemma (i.e., three equally conflicting requirements): high signal-to-noise ratio, thermal stability, and write-ability [2, 4].

PMR is the reason these so-called $L1_0$ ordered alloys are extremely technologically important and heavily researched, although $L1_0$ alloys also have applications in permanent magnets [5] and spintronic technologies [6].

By far, one of the most researched $L1_0$ ordered alloys is FePt because of its superior properties and compatibility with current industrial processes. Therefore, FePt thin films and nanostructures will be given special attention when focus is needed on any aspect of the $L1_0$ ordered alloys. However, this does not diminish the importance of other $L1_0$ alloys (including those yet to be discovered) for each exhibits unique properties and may have its own niche.

There have been many thousands of publications on PMR and the $L1_0$ ordered alloys, which cannot be even remotely discussed or referenced within a chapter, much less a book. Therefore, a subjective attempt was made to highlight important points and reference the most pertinent resources for further explanation and detail.

3.2 Crystal Structure

Certain binary alloy systems, composed of an adequate stoichiometric ratio of A and B atoms, may order in the CuAu I prototype crystal structure with space group *P4/mmm*, which is most widely known by its Strukturbericht notation as the $L1_0$ structure [7]. This chemical ordering, where the A and B atoms occupy specific sites in the unit cell, typically occurs under favorable conditions during deposition or postdeposition annealing.

For example, take the FePt binary alloy system, which exhibits the highest magnetic anisotropy ($K_u \approx 7 \times 10^7$ erg/cm^3) of the $L1_0$ ordered alloys; its bulk equilibrium phase diagram is shown in Fig. 3.2 [8]. There are many phases labeled, including chemically ordered ones (i.e., γ_1, γ_2, γ_3), but the only one of particular interest here is the $L1_0$ phase shaded in gray that occurs within a range of stoichiometries around a 1:1 Fe:Pt atomic ratio. In addition, the equilibrium temperature is just as important as the stoichiometry, and the $L1_0$ phase may form from a high-temperature γ phase of FePt if it is brought to equilibrium below a critical temperature of $\sim$1300°C for a 1:1 Fe:Pt atomic ratio.

It is noticeable that data are sparse or nonexistent for low temperatures ($<$600°C for the FePt case) in bulk phase diagrams because chemical ordering in bulk samples relies on interdiffusion of the two atomic species. This process requires long equilibrium

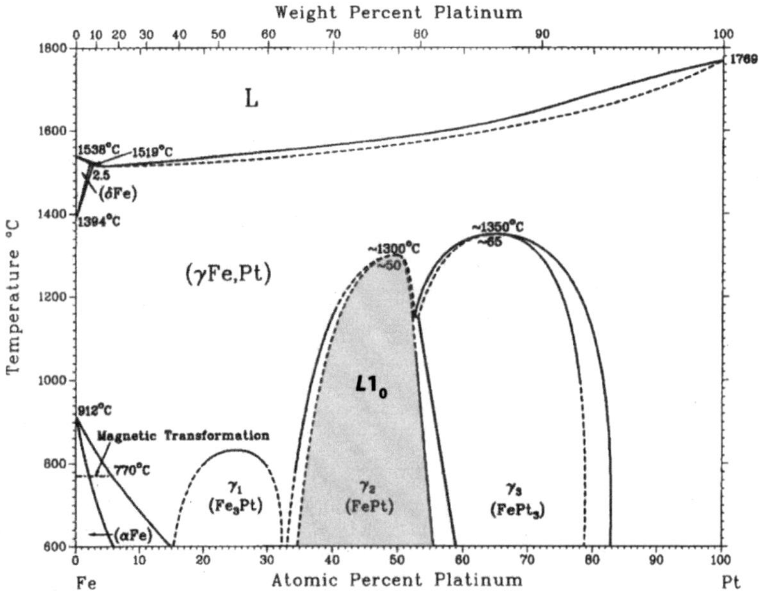

Figure 3.2 The bulk equilibrium phase diagram for FePt with the necessary conditions to achieve the $L1_0$ ordered phase labeled and shaded in gray. From Ref. [8] with permission.

times due to large activation energies, which are extremely slow and prohibitive.

The γ phase of FePt has a disordered face-centered cubic (*fcc*) structure with the space group *Fm-3m*, as shown in Fig. 3.3a, where the Fe and Pt atoms occupy random sites (i.e., one corner and three face centers) in each unit cell. This occurs if conditions are not suitable (e.g., high temperature, incorrect stoichiometry) to promote an ordered structure, which has a lower configuration entropy than the disordered structure.

However, if conditions are favorable, then it is possible that the $L1_0$ ordered face-centered tetragonal (*fct*) structure will form as shown in Fig. 3.3b. This ordering has caused the Fe and Pt atoms to occupy specific positions or sites in the unit cell, thus decreasing the symmetry from a cubic to tetragonal unit cell. Fe atoms occupy the (0, 0, 0) and (1/2, 1/2, 0) sites and the Pt atoms occupy the (0, 1/2, 1/2) and (1/2, 0, 1/2) sites in each unit cell, thus segregating

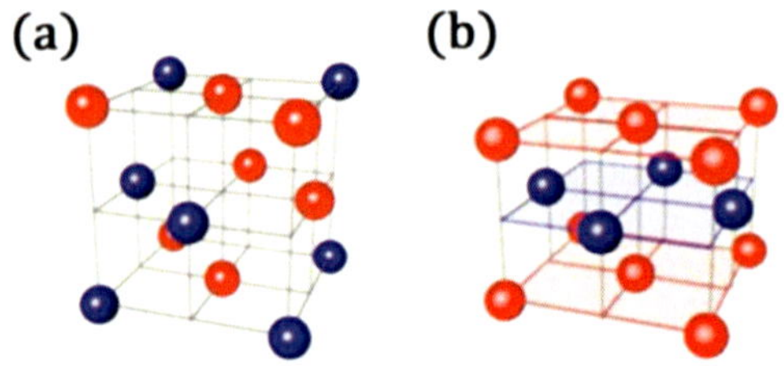

Figure 3.3 (a) The disordered face-centered cubic (*fcc*) unit cell and the (b) $L1_0$ ordered face-centered tetragonal (*fct*) structure.

the Fe and Pt atoms on alternating atomic planes along the c axis or (001) direction. This segregation causes a lattice distortion and hence a change from an *fcc* unit cell in which all sides are equal ($a = b = c$) to an *fct* structure in which the c axis contracts and the a axis expands ($a = b \neq c$), thus resulting in a c/a ratio less than unity. The lattice parameters and c/a ratios for select bulk $L1_0$ ordered alloys are tabulated in Table 3.1 [9].

There is no such *fct* unit cell but in fact a primitive tetragonal unit cell with a basis: the Fe atom at (0,0,0) and the Pt atom at (1/2, 0, 1/2) are both associated with the lattice point at (0,0,0). The base of one of these primitive tetragonal unit cells is formed by the Pt (blue) atoms in Fig. 3.3b.

Table 3.1 Bulk lattice parameters and c/a ratios for select $L1_0$ ordered binary alloys [9]

Binary alloy	a (Å)	c (Å)	c/a	Reference
FePt	3.85	3.71	0.96	[10]
FePd	3.85	3.72	0.97	[11]
CoPt	3.80	3.70	0.97	[10]
CoPd	4.07	3.62	0.89	[12]
NiPt	3.82	3.59	0.94	[13]
MnAl	3.94	3.58	0.91	[14]
MnPt	4.00	3.67	0.92	[15]
MnPd	4.07	3.59	0.88	[16]
MnNi	3.72	3.52	0.95	[16]
CrPt	3.82	3.81	0.99	[17, 18]

Naturally, in the case of these $L1_0$ ordered binary alloys, the two different atoms comprising the binary alloy may occupy either a correct or incorrect lattice site within the unit cell. Again, this is shown in Fig. 3.3 for (a) a completely disordered structure where exactly 50% of the atoms are on the correct sites and (b) a perfectly ordered structure where all of the atoms are on the correct sites. This is termed chemical ordering, and it can be calculated for each individual unit cell, thus allowing the amount of chemical order to be quantified for the entire crystal.

3.3 Long-Range Chemical Order

Chemical order can be simply defined as specific atoms occupying specific lattice sites within a crystalline structure. Short-range chemical order, where an atom on a lattice site influences atoms on nearby lattice sites, is prevalent in many materials systems but only with a range of approximately 1 nm or less. For chemical ordering to influence the properties of the binary alloy, it must possess long-range chemical ordering, which exists over many interatomic distances (i.e., the entire crystal).

3.3.1 *Long-Range Chemical Ordering Parameter*

It is possible to quantify the long-range chemical order within a crystal through its ordering parameter S [19]. Let us revisit and generalize the fully disordered and perfectly ordered crystal structures (unit cells) shown in Fig. 3.3 such that any combination between them (i.e., partial chemical order) is possible.

There are two types of atoms (A and B) that can occupy two types of lattice sites (α and β) within the ordered structure (unit cell). The composition of the crystal must be such that the atom fractions equal unity

$$x_A + x_B = 1, \tag{3.1}$$

and the lattice site fractions equal unity as well

$$y_\alpha + y_\beta = 1. \tag{3.2}$$

Therefore, the total number of atoms N in the crystal will determine the total number of each kind of atom (A and B) and lattice site (α and β):

$$\begin{aligned} N_A &= x_A N \\ N_B &= x_B N \\ N_\alpha &= y_\alpha N \\ N_\beta &= y_\beta N \end{aligned} \tag{3.3}$$

To determine which lattice site an atom is occupying, four parameters must be introduced:

$$\begin{aligned} r_\alpha &: \text{fraction of } \alpha\text{-sites occupied by } A \text{ (correct) atoms} \\ w_\alpha &: \text{fraction of } \alpha\text{-sites occupied by } B \text{ (incorrect) atoms} \\ r_\beta &: \text{fraction of } \beta\text{-sites occupied by } B \text{ (correct) atoms} \\ w_\beta &: \text{fraction of } \beta\text{-sites occupied by } A \text{ (incorrect) atoms} \end{aligned} \tag{3.4}$$

where the following relations must be true:

$$\begin{aligned} r_\alpha + w_\alpha &= 1 \\ r_\beta + w_\beta &= 1 \end{aligned} \tag{3.5}$$

Furthermore, the fraction of A (B) atoms must equal the fraction of lattice sites occupied by those A (B) atoms:

$$\begin{aligned} x_A &= y_\alpha r_\alpha + y_\beta w_\beta \\ x_B &= y_\beta r_\beta + y_\alpha w_\alpha \end{aligned} \tag{3.6}$$

The long-range chemical ordering parameter S can now be defined such that it is linearly proportional to the total fraction of lattice sites occupied by the correct atoms (i.e., $r_\alpha + r_\beta$):

$$S = m(r_\alpha + r_\beta) + b \tag{3.7}$$

Let $S = 0$ for a completely disordered (random) arrangement and $S = 1$ for a completely ordered arrangement (see Fig. 3.3) with an equiatomic composition to yield

$$\begin{aligned} 0 &= m(0.5 + 0.5) + b \\ 1 &= m(1 + 1) + b \end{aligned} \tag{3.8}$$

Solving these two equations simultaneously yields $m = 1$ and $b = -1$, and substituting these values into Eq. 3.7 and using the relations expressed in Eq. 3.5 yields

$$S = r_\alpha + r_\beta - 1 = r_\beta - w_\alpha = r_\alpha - w_\beta \tag{3.9}$$

A convenient expression for interpreting the value of S can be obtained by combining Eqs. 3.6 and 3.9:

$$S = \frac{r_\alpha - x_A}{y_\beta} = \frac{r_\beta - x_B}{y_\alpha} \quad (3.10)$$

Therefore, S reaches its maximum value $S_{\max}$ only when a perfect stoichiometric composition is achieved ($x_A = x_B = 0.5$), since 50% of the lattice sites are for A atoms and 50% for B atoms ($y_\alpha = y_\beta = 0.5$) in an $L1_0$ ordered structure. This can be seen if all of the atoms are on the correct sites ($r_\alpha = r_\beta = 1$), leading to a value of $S = 1$. However, if only 50% of the atoms are on the correct sites ($r_\alpha = r_\beta = 0.5$), then $S = 0$. Of course, any value of S between 0 and 1 (partial chemical order) is attainable with $S_{\max} = 1 - 2\Delta x$, where Δx is the deviation from equiatomic composition ($x_A = x_B = 0.5$).

3.3.2 *Direct Measurement of* S

It is possible to experimentally measure the long-range chemical ordering parameter S of a crystal through direct methods rather than inferring the value of S through structure–property correlations. Two direct methods exist, X-ray diffraction (XRD) and transmission electron microscopy (TEM). XRD is more widely used because it is nondestructive and averages over a large volume of the crystal, whereas TEM is a destructive technique and is only a local probe. Plus, XRD is a relatively less time-consuming technique as compared to TEM. Therefore, the discussion here will be limited to XRD, but the reader is encouraged to seek out literature for the verification of chemical order using TEM, which is especially useful for nanoparticles [9, 20–22].

The XRD pattern of a crystal will differ depending on how the A and B atoms are arranged, whether in a random disordered *fcc* structure (Fig. 3.3a) or in an $L1_0$ ordered *fct* structure (Fig. 3.3b). The diffraction line intensities for certain Miller indices (*hkl* reflections) of the crystal will change depending on the occupation of lattice sites by particular atoms and can be demonstrated by calculating the structure factor F_{hkl} for two extreme cases (i.e., $S = 0$ and $S > 0$).

The structure factor F_{hkl} involves a summation over each and every atomic position within the unit cell and can be written as summations over the α and β lattice sites using average scattering factors f_A and f_B for the A and B atoms, along with the position (x_n, y_n, z_n) of each atom:

$$F_{\text{hkl}} = \sum_{\alpha} (r_\alpha f_A + w_\alpha f_B)\, e^{2\pi i(hx_n+ky_n+lz_n)} + \sum_{\beta} \left(r_\beta f_B + w_\beta f_A\right) e^{2\pi i(hx_n+ky_n+lz_n)} \tag{3.11}$$

3.3.2.1 Case 1: No chemical order ($S = 0$)

Take a completely disordered *fcc* unit cell ($S = 0$), as shown in Fig. 3.3a, where the A and B atoms have an equal probability of being on an α or β lattice site (i.e., $r_\alpha = r_\beta = w_\alpha = w_\beta = 0.5$). This results in α and β lattice sites being equivalent to each other and Eq. 3.11 can now be written as a summation over all lattice sites n:

$$F_{\text{hkl}} = \sum_{n} \left(\frac{f_A + f_B}{2}\right) e^{2\pi i(hx_n+ky_n+lz_n)} \tag{3.12}$$

There are four atoms per unit cell with positions of (0, 0, 0), (1/2, 1/2, 0), (0, 1/2, 1/2), and (1/2, 0, 1/2), and Eq. 3.12 becomes

$$F_{\text{hkl}} = \left(\frac{f_A + f_B}{2}\right) \left[1 + e^{i\pi(h+k)} + e^{i\pi(k+l)} + e^{i\pi(h+l)}\right] \tag{3.13}$$

The value of F_{hkl} depends on the *hkl* reflections of interest; quite simply

$$F_{\text{hkl}} = 2\,(f_A + f_B) \tag{3.14}$$

for *hkl* unmixed (i.e., all even or all odd) reflections, such as the (111) and (002) reflections, and

$$F_{\text{hkl}} = 0 \tag{3.15}$$

for *hkl* mixed (i.e., not all even or all odd) reflections, such as the (001) and (110) reflections. Therefore, mixed *hkl* reflections are forbidden, which is similar to the case for an *fcc* unit cell with only one type of atom.

3.3.2.2 Case 2: Chemical order ($S > 0$)

The previous analysis does not apply when there is any amount of chemical ordering, because the lattice sites are no longer equivalent.

It now matters whether an A or B atom is located on an α or β lattice site. Remember that the two α lattice sites are located at (0, 0, 0) and (1/2, 1/2, 0) and the two β lattice sites at (0, 1/2, 1/2) and (1/2, 0, 1/2). The structure factor F_{hkl} for chemical ordering, following Eq. 3.11, may be written as

$$F_{\text{hkl}} = (r_\alpha f_A + w_\alpha f_B)\left[1 + e^{i\pi(h+k)}\right] + \left(r_\beta f_B + w_\beta f_A\right)\left[e^{i\pi(k+l)} + e^{i\pi(h+l)}\right] \quad (3.16)$$

This equation may be simplified into three different scenarios by using definitions from Section 3.3.1.

$$\begin{aligned} F_{\text{hkl}} &= 2\,(r_\alpha f_A + w_\alpha f_B) + 2\,\left(r_\beta f_B + w_\beta f_A\right) \\ &= 4\,(x_A f_A + x_B f_B) \end{aligned} \quad (3.17)$$

for hkl unmixed, for example, (002) and (111).

$$\begin{aligned} F_{\text{hkl}} &= 2\,(r_\alpha f_A + w_\alpha f_B) - 2\,\left(r_\beta f_B + w_\beta f_A\right) \\ &= 2S\,(f_A - f_B) \end{aligned} \quad (3.18)$$

for hkl mixed with $h + k =$ even and $k + l =$ odd, for example, (001) and (110).

$$F_{\text{hkl}} = 0 \quad (3.19)$$

for hkl mixed with $h + k =$ odd, for example, (011) and (121).

These various hkl reflections can be grouped into one of two categories, fundamental or superstructure reflections. Fundamental reflections are independent of chemical order (i.e., F_{hkl} is independent of S) and therefore appear in diffraction scans of both ordered ($S > 0$) and disordered ($S = 0$) alloys. This is clearly seen by inspecting Eqs. 3.14 and 3.17, which are in fact independent of S and equivalent when $x_A = x_B = 0.5$.

Superstructure reflections, on the other hand, are dependent on the amount of chemical order (i.e., F_{hkl} is dependent on S) and therefore appear only in diffraction scans if some degree of chemical ordering exists ($S > 0$).

Take for example Eq. 3.18. If there is no chemical order ($S = 0$), then Eqs. 3.15, 3.18, and 3.19 are all equivalent and no superstructure reflections appear in the diffraction scans. Furthermore, this is evident if we qualitatively revisit X-ray scattering and the Bragg law [19, 23] along with Fig. 3.3.

Suppose that the Bragg law is satisfied for the (001) superstructure reflection. It then follows that the X-rays from the top and bottom planes of the unit cell scatter in phase, but the X-rays from the middle plane scatter out of phase in relation to the X-rays scattered from the top and bottom planes. Therefore, complete destructive interference results for a completely disordered structure (see Fig. 3.3a), since the structure factors F_{hkl} for each plane of atoms are equivalent, and no (001) superstructure reflection results. However, if there is any amount of chemical ordering (see Fig. 3.3b), then the structure factor F_{hkl} for the middle plane differs from the top and bottom planes (i.e., they contain different atoms), and therefore the (001) superstructure reflection appears indicating chemical order.

3.3.3 *Experimental Quantification of* S

The long-range chemical ordering parameter S can now be calculated by comparing the integrated area of a superstructure reflection (e.g., (001)) to that of a fundamental reflection (e.g., (002)). This is because the intensity of a superstructure reflection is proportional to S^2 (due to the structure factor F_{hkl}), while fundamental reflections are independent of S. The integrated area A_{hkl} of an XRD reflection may be theoretically calculated and written as [9, 19]

$$A_{\mathrm{hkl}} = C_0(LP)F_{\mathrm{hkl}}F^*_{\mathrm{hkl}}e^{-2M}. \tag{3.20}$$

This equation not only contains the structure factor F_{hkl} and its complex conjugate F^*_{hkl}, as previously discussed, but also a number of other correction factors. C_0 is a constant that is dependent on incident intensity and other factors (e.g., absorption correction). The Lorentz factor L corrects for the trigonometric factors that influence the intensity of an X-ray diffracted beam and may be written as [9, 23]

$$L = \frac{1}{\sin\theta\ \sin 2\theta} \quad \text{or} \quad \frac{1}{\sin\theta\ \cos\theta} \tag{3.21}$$

for polycrystalline and epitaxial thin films, respectively, where θ is the diffraction angle.

The polarization factor P corrects for the polarization of the diffracted X-ray beam, as X-rays become partially polarized upon scattering at any angle, and may be written as [9]

$$P = P_0 + \left[(1 - P_0) \cos^2(2\theta) \right], \tag{3.22}$$

where P_0 is the extent of polarization of the incident X-ray beam. Most laboratory X-ray sources are unpolarized ($P_0 = 0.5$), and thus Eq. 3.22 can be written in the more recognized form of

$$P = \frac{1}{2} \left[1 + \cos^2(2\theta) \right]. \tag{3.23}$$

The Debye–Waller factor e^{-2M} is a temperature correction factor that accounts for the intensity decrease of the diffracted beam due to thermal vibrations of the atoms on the lattice. It may be written as [19]

$$2M = \frac{12\, h^2\, T}{m\, k_B\, \Theta_D^2} \left[\Phi(x) + \frac{x}{4} \right] \left(\frac{\sin\theta}{\lambda} \right)^2, \tag{3.24}$$

where h is Planck's constant, T is the temperature, m is the atom's mass, k_B is Boltzmann's constant, Θ_D is the Debye temperature of the atom, λ is the X-ray wavelength, and

$$\left[\Phi(x) + \frac{x}{4} \right] = 1 + \frac{x^2}{36} - \frac{x^4}{3600} + \dots \tag{3.25}$$

where $x = \Theta_D / T$.

One more correction must be applied to the structure factors discussed previously, because the normally tabulated atomic scattering factors f_0 assume that the scattering power is similar to a free electron, that is, the electronic binding energy is small compared to the X-ray photon energy. However, this is not the case because most of the electrons in a periodic lattice are bound and not free. The atomic scattering factor f will now include a real part dispersion correction $\Delta f'$ and an imaginary part dispersion correction $\Delta f''$, such that [24]

$$f = f_0 + \Delta f' + i \Delta f''. \tag{3.26}$$

The structure factor F_{hkl} for a fundamental reflection (Eq. 3.17), incorporating the Debye–Waller factor, may now be written as

$$\begin{aligned} F_{\mathrm{hkl}} = 4 \big[& x_A \left(f_A + \Delta f'_A + i \Delta f''_A \right) e^{-M_A} \\ & + x_B \left(f_B + \Delta f'_B + i \Delta f''_B \right) e^{-M_B} \big] \end{aligned} \tag{3.27}$$

with its complex conjugate F^*_{hkl} expressed as

$$F^*_{\text{hkl}} = 4\left[x_A\left(f_A + \Delta f'_A - i\Delta f''_A\right)e^{-M_A} \right.$$
$$\left. + x_B\left(f_B + \Delta f'_B - i\Delta f''_B\right)e^{-M_B}\right]. \quad (3.28)$$

Multiplying these two equations together yields the square magnitude of the structure factor F^2_{hkl} for a fundamental reflection

$$F_{\text{hkl}}F^*_{\text{hkl}} = 16\left[\left(x_A\left(f_A + \Delta f'_A\right)e^{-M_A} + x_B\left(f_B + \Delta f'_B\right)e^{-M_B}\right)^2 \right.$$
$$\left. + \left(x_A\Delta f''_A e^{-M_A} + x_B\Delta f''_B e^{-M_B}\right)^2\right]. \quad (3.29)$$

The square magnitude of the structure factor F^2_{hkl} for a superstructure may be written in a similar fashion as

$$F_{\text{hkl}}F^*_{\text{hkl}} = 4S^2\left[\left(\left(f_A + \Delta f'_A\right)e^{-M_A} - \left(f_B + \Delta f'_B\right)e^{-M_B}\right)^2 \right.$$
$$\left. + \left(\Delta f''_A e^{-M_A} - \Delta f''_B e^{-M_B}\right)^2\right]. \quad (3.30)$$

Since $F_{\text{hkl}}F^*_{\text{hkl}}$ of a superstructure reflection is dependent on S^2, a ratio of the superstructure to fundamental reflections allows a solution of S to be found by simply comparing the measured XRD integrated intensities I_{hkl} to the theoretical values A_{hkl} from Eq. 3.20:

$$\frac{I_{001}}{I_{002}} = \frac{(LPe^{-2M})_{001}F_{001}F^*_{001}}{(LPe^{-2M})_{002}F_{002}F^*_{002}} \quad (3.31)$$

The error in the long-range chemical ordering parameter S can be calculated by taking the standard deviation between two sets of superstructure/fundamental reflections using Eq. 3.31, for instance, the (001)/(002) and (003)/(002) reflections, which typically yield error values less than 0.1.

3.4 Magnetic Anisotropy and Other Properties

The structural anisotropy of these $L1_0$ ordered alloys (i.e., long-range chemical order) logically gives rise to their strong magnetic anisotropy. In particular, the magnetocrystalline anisotropy (MCA) is the dominant factor, since the spin moments of the electrons are coupled to the orbital moments from the crystal lattice through

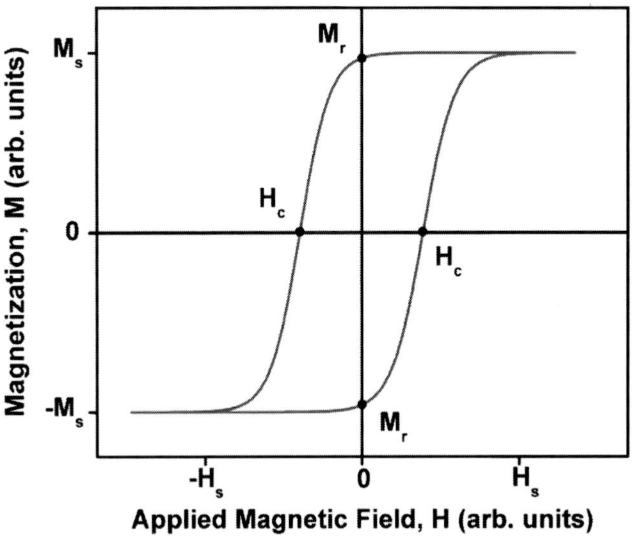

Figure 3.4 Hysteresis or **M**(**H**) loop with important parameters labeled: saturation magnetization $\mathbf{M}_s$, coercive field $\mathbf{H}_C$, saturation field $\mathbf{H}_s$, and remanent magnetization $\mathbf{M}_r$.

spin–orbit (SO) interactions [25]. These SO interactions describe the microscopic origin of the MCA and are largest for FePt because of the large SO coupling for Pt combined with the spin polarization of the Pt 5*d* bands [26].

The magnetic properties that are of interest to the $L1_0$ ordered alloys include not only the uniaxial magnetic anisotropy energy density K_u but also the saturation magnetization $\mathbf{M}_s$, coercive field $\mathbf{H}_C$, saturation field $\mathbf{H}_s$, anisotropy field $\mathbf{H}_K$, and remanent magnetization $\mathbf{M}_r$. These properties can be experimentally obtained from a hysteresis or **M**(**H**) loop, where the magnetization **M** is measured as a function of an applied magnetic field **H** at a fixed temperature, an example of which is shown in Fig. 3.4.

The anisotropy field $\mathbf{H}_K$ and the uniaxial magnetic anisotropy energy density K_u can be calculated using the relations [27]

$$K_u = K_{eff} + 2\pi M_s^2$$

$$\mathbf{H}_K = \frac{2K_u}{\mathbf{M}_s}, \tag{3.32}$$

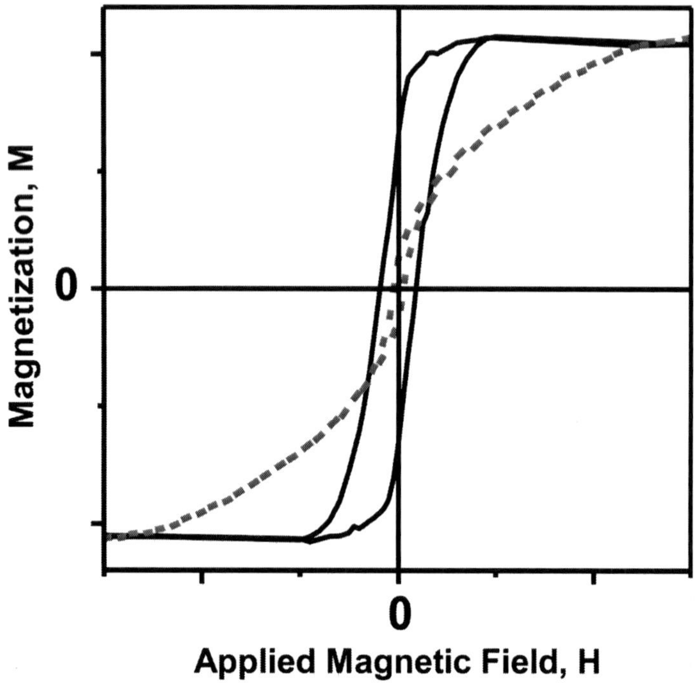

Figure 3.5 Perpendicular (black solid) and in-plane (dotted) hysteresis or **M**(**H**) loops showing the typical characteristics of the easy and hard axis directions, respectively. The easy axis has low $\mathbf{H}_s$, large $\mathbf{H}_C$, and high $\mathbf{M}_r$, whereas the hard axis has large $\mathbf{H}_s$, small $\mathbf{H}_C$, and low $\mathbf{M}_r$.

where K_{eff} is the effective uniaxial anisotropy constant. K_{eff} can be calculated from the area between the **M**(**H**) loops for the easy (out-of-plane direction) and hard (in-plane direction) axes [28] or using a 45° method and high-field extrapolation [29]. Typical in-plane (hard axis) and perpendicular (easy axis) **M**(**H**) loops for the $L1_0$ ordered alloys are shown in Fig. 3.5 with typical magnetic property values listed in Table 3.2.

The magnetic properties of these alloys, especially K_u, strongly affect the theoretical minimum grain size that can hold a magnetic bit for data storage applications discussed previously. Since these magnetic properties are strongly correlated to the $L1_0$ ordered structure, deposition techniques to synthesize these materials in

Table 3.2 Magnetic properties of various PMR media candidates with theoretical minimum spherical grain sizes D_p [3]

Alloy material	K_u (10^7 erg/cm^3)	M_s (emu/cm^3)	H_K (kOe)	D_p (nm)
Co_3Pt	2	1100	36.4	8.5
MnAl	1.7	560	60.7	7.1
FePd	1.8	1100	32.7	9.3
CoPt	4.9	800	122.5	4.9
FePt	7	1140	122.8	4.4
$SmCo_5$	20	910	439.6	2.9

thin-film and nanostructured forms become a critical tool for tailoring their properties.

3.5 Material Synthesis

These $L1_0$ ordered alloys, as with many other materials systems, were first studied in their bulk forms, leading to equilibrium phase diagrams, such as those shown for FePt in Fig. 3.2. Only recently has deposition technology improved to the extent that unprecedented control of the crystal structure could be demonstrated in thin-film and nanostructured forms. This control of the crystal structure at the atomic level has allowed precise and tailored deposition of thin films and nanostructures for both fundamental studies and technological applications.

For instance, the temperatures required to achieve $L1_0$ chemical order are far lower than those presented in bulk phase diagrams, because the nonequilibrium kinetics or dynamics of these deposition techniques are strongly governed by the adatom mobility at the growing film interface (i.e., low activation energies) as opposed to bulk diffusion (i.e., high activation energies) [30]. This is extremely attractive for technological applications, as lower processing temperatures decrease the time and cost associated with commercial production. In addition, these thin-film deposition techniques have facilitated the growth of single crystals, which is difficult to do in bulk form without an external magnetic field [31, 32] or an elastic stress [33].

3.5.1 *Codeposition Techniques for Thin Films*

The predominant thin-film codeposition techniques for the $L1_0$ ordered alloys are molecular beam epitaxy (MBE) [34, 35] and sputter deposition [36, 37]. MBE is the preferred deposition method on the research side of investigating the $L1_0$ ordered alloys for various reasons. Firstly, the background pressure during deposition is typically in the ultrahigh vacuum (UHV) regime (i.e., $<10^{-9}$ Torr), which limits the incorporation of impurities or defects in the thin film if suitable growth rates are used ($>$0.05 Å/s). Secondly, in situ characterization techniques, such as reflection high-energy electron diffraction (RHEED) [38], low-energy electron diffraction (LEED) [39], and X-ray photoelectron spectroscopy (XPS) [40], can be used continuously to monitor film deposition from beginning to end. However, on the industrial and commercial sides, sputter deposition is the preferred deposition method.

Both techniques can favor the preferential growth of the $L1_0$ (001) ordered structure if a suitable single-crystalline substrate is chosen, along with proper deposition conditions (e.g., substrate temperature, growth rate). This is demonstrated in Fig. 3.6 for 40 nm thick FePd (001) thin films cosputter-deposited onto single-crystalline MgO (100) substrates by only varying the deposition (substrate) temperature [41]. Various microstructural and chemical ordering parameters (using the aforementioned analysis) for these FePd films are presented in Table 3.3, demonstrating the need for precise control of experimental parameters.

Table 3.3 Microstructural parameters extracted from XRD analysis of 40 nm thick FePd films grown at different deposition (substrate) temperatures [41]

Deposition temperature (°C)	a (Å)	c (Å)	c/a	S_{order}
400	3.852	3.688	0.9574	0.80
450	3.851	3.685	0.9569	0.75
500	3.855	3.696	0.9588	0.77
550	3.847	3.709	0.9641	0.58
600	3.850	3.720	0.9662	0.62

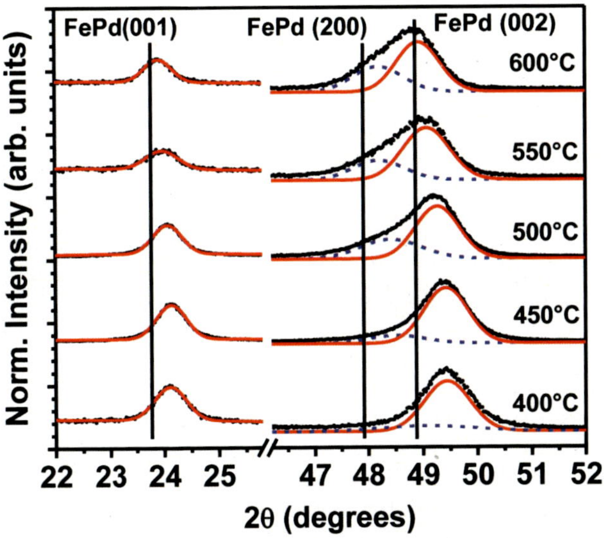

Figure 3.6 Symmetric θ–2θ XRD scans showing how the deposition (substrate) temperature affects the $L1_0$ ordering for 40 nm thick FePd films (black data). Gaussian fits for the FePd superstructure and fundamental reflections (red solid lines), as well as the disordered *fcc* FePd phase (blue dashed lines), are shown along with their bulk lattice parameters (vertical lines) [41].

Of course, other deposition techniques such as pulsed laser deposition (PLD) [42–45], oblique angle deposition (OAD) or glancing angle deposition (GLAD) [46–49], and electrodeposition [50–52] have been used extensively as well.

Postdeposition annealing treatments, either in conventional or in rapid thermal annealing (RTA) [53] ovens, can also be used to increase the chemical order and correlated properties of the as-deposited films [54].

In addition to simultaneous codeposition of the two atomic species, there are other codeposition techniques to achieve and influence the chemical order in thin films and their correlated properties. One of those methods is artificially synthesizing the $L1_0$ structure by depositing alternating monolayers of the two atomic species, which can be achieved by atomic-scale multilayer sputtering

(AMS) [55, 56] or MBE [57] at temperatures much lower than required by simultaneous codeposition techniques.

Another method is to deposit multilayers (i.e., many monolayers) of the two elements followed by a postdeposition annealing treatment (conventional or RTA) [58] or ion-beam irradiation [59, 60]. Recently, Ga^+ irradiation–driven modification of magnetic properties has been demonstrated in Pt/Co/Pt ultrathin films, thus offering promise to design high-resolution nanostructures with tailored magnetic properties [61].

3.5.2 *Incorporation of Additional Elements*

In addition to these various deposition techniques and annealing treatments, it is also possible to incorporate other materials into the $L1_0$ alloys in order to alter their structural and correlated magnetic properties. One method is to create a ternary alloy by incorporating an additive element into the $L1_0$ ordered alloy. Elemental additives such as Ni [62], Cr [63], Ag and Au [64, 65], Mg, V, Mn, and B [66], Nb [67], and N [68] have been investigated and show promise in tailoring these FePt $L1_0$ alloys for magnetic media. One of the most promising additives, though, is Cu [55, 64, 69–71] due to its high diffusivity, good thermal conduction, and ability to accelerate the $L1_0$ ordering kinetics, thus lowering processing temperatures.

Incorporating other elemental thin-film materials above (overlayers or capping layers) and below (underlayers or seed layers) the $L1_0$ alloy thin film has shown promise as well to enhance chemical ordering and associated magnetic properties. Seed layers of Ag [72], Pt [73], Cr [74], Cr and Pt [27], RuAl [75], CrRu and Pt [76], and Fe and Pt [77] have been extensively researched, as well as others not mentioned here. The use of different overlayers or capping layers, such as Pt [78], Pd, or MgO [79], show influence on the magnetic anisotropy of FePt and FePd thin films.

Nanocomposites offer another avenue of combining other materials with the $L1_0$ alloys, in which small $L1_0$ ordered grains are dispersed within a nonmagnetic matrix. FePt–Ag [80], FePt:SiO_2 [81], and FePt:B_2O_3 [82] nanocomposites have been investigated by annealing as-deposited FePt/nonmagnetic multilayers.

3.5.3 *Nanostructures*

Nanostructures have become increasingly important because novel properties begin to emerge as one or more length scales shrink (i.e., approach some characteristic/physical length scale, typically in the nanometer regime). There are various nanofabrication methods [83] that can create magnetic nanostructures from the micron- to the nanoscale from a top-down or bottom-up approach. These nanostructures are created through the standard lithographic process: resist coating, exposing, developing, and etching (or depositing followed by lift-off), as shown in Fig. 3.7. The attainable size and spacing of the patterned nanostructures is highly dependent on the radiation wavelength used during the exposure process and hence is used to name the various mainstream techniques: electron beam, X-ray, laser interference, scanning probe, and nanoimprint lithographies.

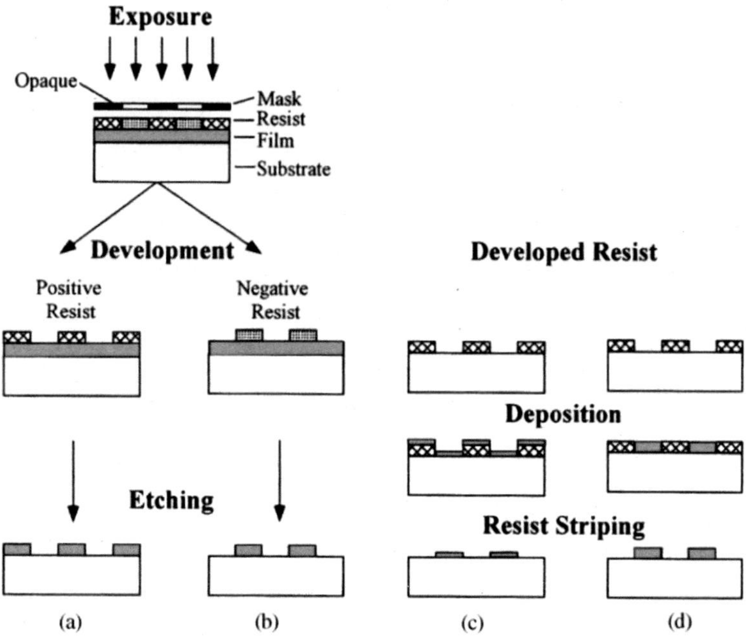

Figure 3.7 Schematics of lithography processes for (a) positive and (b) negative resists in conjunction with (a, b) etching, (c) lift-off, and (d) electrodeposition. From Ref. [83] with permission.

Each lithography technique has its limitations: minimum feature size, maximum patterning area, and time/cost. Recently, nanoimprint lithography has been able to produce a 4 mm^2 $L1_0$ ordered FePt nanopattern composed of 60 nm dots with a periodicity of 150 nm [84], which has grand implications for developing next-generation novel recording technologies, such as bit-patterned media (BPM) [85, 86].

Chemical synthesis of nanoparticles is also prevalent in the literature, with the Fe–Pt system leading the way for magnetic data storage applications [87]. These methods typically include solution chemistry and an arrested precipitation approach followed by deposition onto a substrate [88]. Gas phase–based approaches, such as ion-beam-induced chemical vapor deposition [89], have also been explored to overcome challenges met from these solution-based methods.

3.6 Conclusion

Nanostructured $L1_0$ ordered thin films, most likely via nanolithography patterning techniques described previously, have the potential to keep pushing the limits of magnetic recording technology for HDDs. PMR has already begun to replace LMR in commercial HDDs, with the potential for $L1_0$ ordered alloys to be incorporated into PMR and other technologies to further increase the areal density.

The areal density is fundamentally limited by the grain size of the media, which must remain stable as the grains become smaller (i.e., the media must have a large magnetic anisotropy energy density K_u). $L1_0$ ordered alloys have some of the highest values of K_u, which make them great candidates to increase the areal density. However, a larger K_u also means that a larger write field is needed (much larger than the ~24 kOe that current write heads are able to deliver).

This problem, part of the trilemma of magnetic recording, can be solved using a number of technologies still in development. Write-assist technologies, such as heat-assisted magnetic recording (HAMR) [90] and microwave-assisted magnetic recording (MAMR) [91], pump in energy to make the magnetization easier to switch (i.e., at a lower magnetic field). Another option with write-assist is to build it into the media. Exchange-coupled composite (ECC) media [92] uses a soft magnetic layer to assist in the switching of a hard magnetic layer when an applied field is present through domain wall–assisted switching [93].

The media can also be changed further so that instead of a quasi-homogenous medium where multiple magnetic grains define one magnetic bit, one grain will define one magnetic bit. This is the idea behind BPM [94]. In any case and whichever method prevails, $L1_0$ ordered alloys offer the necessary structural and magnetic properties to keep pushing the superparamagnetic limit in the quest for larger areal densities in magnetic data storage media.

Acknowledgments

I would like to acknowledge my collaborators: R. A. Lukaszew, C. Clavero, K. Yang, B. Wincheski, Y. Choi, D. Haskel, J. M. García-Martín, and A. Cebollada. This work was supported by the American Chemical Society (PRF-41319-AC10), Department of Defense-US Army Research Office (W911NF-11-1-0209 and W911NF-11-1-0133), National Science Foundation (DMR-0355171 and CREST HRD-1036494), Research Corporation (Cottrell Scholar Award), and Virginia Space Grant Consortium.

References

1. E. Grochowski and R. D. Halem, Technological impact of magnetic hard disk drives on storage systems, *IBM Syst. J.*, **42**, 338–346 (2003).
2. H. J. Richter, The transition from longitudinal to perpendicular recording, *J. Phys. D.: Appl. Phys.*, **40**, R149–R177 (2007).
3. D. Weller and T. McDaniel, Media for extremely high density recording, in *Advanced Magnetic Nanostructures*, 295–324, D. Sellmyer and R. Skomski, Eds. New York: Springer, 2006.
4. H. J. Richter, Density limits imposed by the microstructure of magnetic recording media, *J. Magn. Magn. Mater.*, **321**, 467–476 (2009).

5. O. Gutfleishch, M. A. Willard, E. Brück, C. H. Chen, S. G. Sankar, and J. P. Liu, Magnetic materials and devices for the 21st century: stronger, lighter, and more energy efficient, *Adv. Mater.*, **23**, 812–842 (2010).
6. S. A. Wolf, J. Lu, M. R. Stan, E. Chen, and D. M. Treger, The promise of nanomagnetics and spintronics for future logic and universal memory, *Proc. IEEE*, **98**, 2155–2168 (2010).
7. D. E. Laughlin, K. Srinivasan, M. Tanase, and L. Wang, Crystallographic aspects of $L1_0$ magnetic materials, *Scripta Mater.*, **53**, 383–388 (2005).
8. T. B. Massalski, H. Okamoto, P. R. Subramanian, and L., Kacprzak, Eds., *Binary Alloy Phase Diagrams,* 2nd ed., 1751–1752, Materials Park, OH: ASM International, 1990.
9. A. Cebollada, R. F. C. Farrow, and M. F. Toney, Structure and magnetic properties of chemically ordered magnetic binary alloys in thin film form, in *Magnetic Nanostructures*, 93–122, H. S. Nalwa, Ed. Stevenson Ranch, CA: American Scientific, 2002.
10. P. Villars and L. D. Culvert, *Pearson's Handbook of Crystallographic Data for Intermetallic Phases*, Materials Park, OH: ASM International, 1991.
11. W. B. Pearson, *Lattice Spacings and Structures of Metals and Alloys*, New York: Pergamon, 1967.
12. Y. Matsuo, Ordered alloys in the cobalt-palladium system, *J. Phys. Soc. Japan*, **32**, 972–978 (1972).
13. M. Hansen, *Constitution of Binary Alloys,* 2nd ed., New York: McGraw-Hill, 1958.
14. H. Kono, On the ferromagnetic phase in manganese-aluminum system, *J. Phys. Soc. Jpn.*, **13**, 1444–1451 (1958).
15. H. H. Landolt, *Landolt-Bornstein, New Series,* Vol. 19a, Berlin: Springer-Verlag, 1986.
16. J. S. Kasper and J. S. Kouvel, The antiferromagnetic structure of NiMn, *Phys. Chem. Solids*, **11**, 231–238 (1959).
17. M. J. Besnus and A. J. P. Meyer, Magnetic properties of the ordered and disordered $CrPt_3$ and CrPt phases, *Phys. Status Solidi B*, **58**, 533–542 (1973).
18. R. Zhang, R. Skomski, X. Li, Z. Li, P. Manchanda, A. Kashyap, R. D. Kirby, S.-H. Liou, and D. J. Sellmyer, $L1_0$ CrPt phase formation and magnetic properties, *J. Appl. Phys.*, **111**, 07D720 (2012).
19. B. E. Warren, *X-ray Diffraction*, Reading, MA: Addison-Wesley, 1969.
20. R. F. C. Farrow, D. Weller, R. F. Marks, M. F. Toney, D. J. Smith, and M. R. McCartney, Magnetic anisotropy and microstructure in molecular beam epitaxial FePt(110)/MgO(110), *J. Appl. Phys.*, **84**, 934–939 (1998).

21. C.-B. Rong, D. Li, V. Nandwana, N. Poudyal, Y. Ding, Z. L. Wang, H. Zeng, and J. P. Liu, Size-dependent chemical and magnetic ordering in $L1_0$-FePt nanoparticles, *Adv. Mater.*, **18**, 2984–2988 (2006).
22. S. Sun, C. B. Murray, D. Weller, L. Folks, and A. Moser, Monodisperse FePt nanoparticles and ferromagnetic FePt nanocrystal superlattices, *Science*, **287**, 1989–1992 (2000).
23. B. D. Cullity and S. R. Stock, *Elements of x-ray diffraction*, 3rd ed., Upper Saddle River, NJ: Prentice-Hall, 2001.
24. C. H. MacGillavry, G. D. Rieck, and K. Lonsdale, *International Tables for X-Ray Crystallography*, Vol. 3, 3rd ed., Birmingham, UK: Kynoch Press, 1969.
25. J. A. C. Bland and B. Heinrich, Eds., *Ultrathin Magnetic Structures I: an Introduction to the Electronic, Magnetic, and Structural Properties*, Berlin: Springer-Verlag, 1994.
26. G. H. O. Daalderop, P. J. Kelly, and M. F. H. Schuurmans, Magnetocrystalline anisotropy and orbital moments in transition-metal compounds, *Phys. Rev. B*, **44**, 12054–12057 (1991).
27. J.-U. Thiele, L. Folks, M. F. Toney, and D. K. Weller, Perpendicular magnetic anisotropy and magnetic domain structure in sputtered epitaxial FePt (001) $L1_0$ films, *J. Appl. Phys.*, **84**, 5686–5692 (1998).
28. M. T. Johnson, P. J. H. Bloemen, F. J. A. den Broeder, and J. J. de Vries, Magnetic anisotropy in metallic multilayers, *Rep. Prog. Phys.*, **59**, 1409–1458 (1996).
29. H. Miyajima, K. Sato, and T. Mizoguchi, Simple analysis of torque measurement of magnetic thin films, *J. Appl. Phys.*, **47**, 4669–4671 (1976).
30. G. R. Harp, D. Weller, T. A. Rabedeau, R. F. C. Farrow, and M. F. Toney, Magneto-optical Kerr spectroscopy of a new chemically ordered alloy: Co_3Pt, *Phys. Rev. Lett.*, **71**, 2493–2496 (1993).
31. P. Eurin and J. Pauleve, Influence of thermomagnetic treatments on the magnetic properties of Co–Pt 50–50 alloy, *IEEE Trans. Magn.*, **5**, 216–219 (1969).
32. K. Tanaka, T. Ichitsubo, M. Amano, M. Koiwa, and K. Watanabe, Formation of mono-variant $L1_0$ structure on ordering of FePd under magnetic fields, *Mater. Trans., JIM*, **41**, 917–922 (2000).
33. T. Ichitsubo, M. Nakamoto, K. Tanaka, and M. Koiwa, Effect of applied stress on fcc-$L1_0$ transformation of FePd single crystal, *Mater. Trans., JIM*, **39**, 24–30 (1998).

34. G. R. Harp, D. Weller, T. A. Rabedeau, R. F. C. Farrow, and R. F. Marks, Magneto optical Kerr spectroscopy as a probe of chemical ordering in CoPt, *Mater. Res. Soc. Proc.*, **313**, 493 (1993).
35. A. Cebollada, D. Weller, J. Sticht, G. R. Harp, R. F. C. Farrow, R. F. Marks, R. Savoy, and J. C. Scott, Enhanced magneto-optical Kerr effect in spontaneously ordered FePt alloys: quantitative agreement between theory and experiment, *Phys. Rev. B*, **50**, 3419–3422 (1994).
36. B. M. Lairson, M. R. Visokay, E. E. Marinero, R. Sinclair, and B. M. Clemens, Epitaxial tetragonal PtCo (001) thin films with perpendicular magnetic anisotropy, *J. Appl. Phys.*, **74**, 1922–1928 (1993).
37. M. R. Visokay and R. Sinclair, Direct formation of ordered CoPt and FePt compound thin films by sputtering, *Appl. Phys. Lett.*, **66**, 1692–1694 (1995).
38. A. Ichimiya and P. I. Cohen, *Reflection High Energy Electron Diffraction*, Cambridge, UK: Cambridge University Press, 2004.
39. M. A. Van Hove, W. H. Weinberg, and C.-M. Chan, *Low-Energy Electron Diffraction: Experiment, Theory, and Surface Structure Determination*, Berlin: Springer-Verlag, 1986.
40. P. Van der Heide, *X-ray Photoelectron Spectroscopy: An Introduction to Principles and Practices*, Hoboken, NJ: John Wiley and Sons, 2012.
41. J. R. Skuza, C. Clavero, K. Yang, B. Wincheski, and R. A. Lukaszew, Microstructural, magnetic anisotropy, and magnetic domain structure correlations in epitaxial FePd thin films with perpendicular magnetic anisotropy, *IEEE Trans. Magn.*, **46**, 1886–1889 (2010).
42. D. B. Chrisey and G. K. Hubler, Eds., *Pulsed Laser Deposition of Thin Films*, New York: John Wiley and Sons, 1994.
43. J. Shen, Z. Gai, and J. Kirschner, Growth and magnetism of metallic thin films and multilayers by pulsed-laser deposition, *Surf. Sci. Rep.*, **52**, 163–218 (2004).
44. G. R. Trichy, J. Narayan, and H. Zhou, $L1_0$ ordered epitaxial FePt (001) thin films on TiN/Si (100) by pulsed laser deposition, *Appl. Phys. Lett.*, **89**, 132502 (2006).
45. P. Schaaf, K. Zhang, C. Lange, A. Holz, M. Weisheit, and S. Fähler, Structure and anisotropy of epitaxial fcc FePt films, *Appl. Surf. Sci.*, **253**, 8107–8110 (2007).
46. K. Robbie and M. J. Brett, Sculptured thin films and glancing angle deposition: growth mechanics and applications, *J. Vac. Sci. Tech. A*, **15**, 1460–1465 (1997).

47. M. M. Hawkeye and M. J. Brett, Glancing angle deposition: Fabrication, properties, and applications of micro- and nanostructured thin films, *J. Vac. Sci. Tech. A*, **25**, 1317–1335 (2007).

48. S. Thomas, S. H. Al-Harthi, I. A. Al-Omari, R. V. Ramanujan, V. Swaminathan, and M. R. Anantharaman, Influence of substrate topography on the growth and magnetic properties of obliquely deposited amorphous nanocolumns of Fe–Ni, *J. Phys. D.: Appl. Phys.*, **42**, 215005 (2009).

49. A. K. Kar, P. Morrow, X.-T. Tang, T. C. Parker, H. Li, J.-Y. Dai, M. Shima, and G.-C. Wang, Epitaxial multilayered Co/Cu ferromagnetic nanocolumns grown by oblique angle deposition, *Nanotechnology*, **18**, 295702 (2007).

50. Y. H. Huang, H. Okumura, G. C. Hadjipanayis, and D. Weller, CoPt and FePt nanowires by electrodeposition, *J. Appl. Phys.*, **91**, 6869–6871 (2002).

51. F. M. F. Rhen, G. Hinds, C. O'Reilly, and J. M. D. Coey, Electrodeposited FePt films, *IEEE Trans. Magn.*, **39**, 2699–2701 (2003).

52. K. Leistner, J. Thomas, H. Schlörb, M. Weisheit, L. Schultz, and S. Fähler, Highly coercive electrodeposited FePt films by post annealing in hydrogen, *Appl. Phys. Lett.*, **85**, 3498–3500 (2004).

53. R. B. Fair, *Rapid Thermal Processing: Science and Technology*, San Diego, CA: Academic Press, 1993.

54. K. Aimuta, K. Nishimura, S. Hashi, and M. Inoue, Fabrication of $L1_0$-FePt thin films by rapid thermal annealing, *IEEE Trans. Magn.*, **41**, 3898–3900 (2005).

55. D. A. Gilbert, L.-W. Wang, T. J. Klemmer, J.-U. Thiele, C.-H. Lai, and K. Liu, Tuning magnetic anisotropy in (001) oriented $L1_0$ $(Fe_{1-x}Cu_x)_{55}Pt_{45}$ films, *Appl. Phys. Lett.*, **102**, 132406 (2013).

56. T. Shima, T. Moriguchi, T. Seki, S. Mitani, and K. Takanashi, Fabrication of $L1_0$ ordered FePt alloy films by monatomic layer sputter deposition, *J. Appl. Phys.*, **93**, 7238–7240 (2003).

57. T. Shima, T. Moriguchi, S. Mitani, and K. Takanashi, Low-temperature fabrication of $L1_0$ ordered FePt alloy by alternate monatomic layer deposition, *Appl. Phys. Lett.*, **80**, 288–290 (2002).

58. T. Y. Lee, D. S. Son, S. H. Lim, and S.-R. Lee, High post-annealing stability in [Pt/Co] multilayers, *J. Appl. Phys.*, **113**, 216102 (2013).

59. S. Bera, B. Satpati, D. K. Goswami, K. Bhattacharjee, P. V. Satyam, and B. N. Dev, Ion-beam induced transformations in nanoscale multilayers: evolution of clusters with preferred length scales, *J. Appl. Phys.*, **99**, 074301 (2006).

60. T. Yokota, M. L. Yan, Y. Xu, L. Gao, R. Zhang, L. Nicholl, L. Yuan, R. Skomski, D. J. Sellmyer, S. H. Liou, C.-H. Lai, C.-H. Yang, and S.-H. Huang, Magnetic properties and $L1_0$ phase formation of FePt films prepared by high current-density ion-beam irradiation and rapid thermal annealing methods, *J. Appl. Phys.*, **97**, 10H306 (2005).

61. A. Maziewski, P. Mazalski, Z. Kurant, M. O. Liedke, J. McCord, J. Fassbender, J. Ferre, A. Mougin, A. Wawro, L. T. Baczewski, A. Rogalev, F. Wilhelm, and T. Gemming, Tailoring of magnetism in Pt/Co/Pt ultrathin films by ion irradiation, *Phys. Rev. B*, **85**, 054427 (2012).

62. J.-U. Thiele, K. R. Coffey, M. F. Toney, J. A. Hedstrom, and A. J. Kellock, Temperature dependent magnetic properties of highly chemically ordered $Fe_{55-x}Ni_xPt_{45}$ $L1_0$ films, *J. Appl. Phys.*, **91**, 6595–6600 (2002).

63. W. Y. Zhang, H. Shima, F. Takano, H. Akinaga, X. Z. Yu, T. Hara, W. Z. Zhang, K. Kimoto, Y. Matsui, and S. Nimori, Enhancement in ordering of $Fe_{50}Pt_{50}$ film caused by Cr and Cu additives, *J. Appl. Phys.*, **106**, 033907 (2009).

64. C. L. Platt, K. W. Wierman, E. B. Svedberg, R. Van de Veerdonk, J. K. Howard, A. G. Roy, and D. E. Laughlin, $L1_0$ ordering and microstructure of FePt thin films with Cu, Ag, and Au additive, *J. Appl. Phys.*, **92**, 6104–6109 (2001).

65. B. Wang, K. Barmak, and T. J. Klemmer, A1 to $L1_0$ transformation in FePt films with ternary alloying additions of Ag and Au, *IEEE Trans. Magn.*, **46**, 1773–1776 (2010).

66. B. Wang, K. Barmak, and T. J. Klemmer, The A1 to $L1_0$ transformation in FePt films with ternary alloying additions of Mg, V, Mn, and B, *J. Appl. Phys.*, **109**, 07B739 (2011).

67. S.-K. Chen, F. T. Yuan, G. S. Chen, and W. C. Chang, Structural and magnetic properties of Fe–Pt–Nb sputtered films, *Phys. B*, **327**, 366–369 (2003).

68. H. Y. Wang, W. H. Mao, X. K. Ma, H. Y. Zhang, Y. B. Chen, Y. J. He, and E. Y. Jiang, Improvement in hard magnetic properties of FePt films by N addition, *J. Appl. Phys.*, **95**, 2564–2568 (2004).

69. C. L. Zha, R. K. Dumas, Y. Y. Fang, V. Bonanni, J. Nogués, and J. Åkerman, Continuously graded anisotropy in single $(Fe_{53}Pt_{47})_{100-x}Cu_x$ films, *App. Phys. Lett.*, **97**, 182504 (2010).

70. B. Wang and K. Barmak, Re-evaluation of the impact of ternary additions of Ni and Cu on the A1 to $L1_0$ transformation in FePt films, *J. Appl. Phys.*, **109**, 123916 (2011).

71. C. Brombacher, H. Schletter, M. Daniel, P. Matthes, N. Jöuhrmann, M. Maret, D. Makarov, M. Hietschold, and M. Albrecht, FePtCu alloy thin

films: Morphology, $L1_0$ chemical ordering, and perpendicular magnetic anisotropy, *J. Appl. Phys.*, **112**, 073912 (2012).

72. Y.-N. Hsu, S. Jeong, D. E. Laughlin, and D. N. Lambeth, Effects of Ag underlayers on the microstructure and magnetic properties of epitaxial FePt thin films, *J. Appl. Phys.*, **89**, 7068–7070 (2001).
73. M. H. Hong, K. Hono, and M. Watanabe, Microstructure of FePt/Pt magnetic thin films with high perpendicular coercivity, *J. Appl. Phys.*, **84**, 4403–4409 (1998).
74. T. Suzuki, K. Harada, N. Honda, and K. Ouchi, Preparation of ordered Fe–Pt thin films for perpendicular magnetic recording media, *J. Magn. Magn. Mater.*, **193**, 85–88 (1999).
75. E. Yang, S. Ratanaphan, D. E. Laughlin, and J.-G. Zhu, Highly ordered FePt $L1_0$ thin films with small grains on RuAl seed layers, *IEEE Trans. Magn.*, **47**, 81–86 (2011).
76. J. S. Chen, Y. Xu, and J. P. Wang, Effect of Pt buffer layer on structural and magnetic properties of FePt thin films, *J. Appl. Phys.*, **93**, 1661–1665 (2003).
77. K. Barmak, J. Kim, L. H. Lewis, K. R. Coffey, M. F. Toney, A. J. Kellock, and J.-U. Thiele, On the relationship of magnetocrystalline anisotropy and stoichiometry in epitaxial $L1_0$ CoPt (001) and FePt (001) thin films, *J. Appl. Phys.*, **98**, 033904 (2005).
78. M. Tsujikawa, A. Hosokawa, and T. Oda, Magnetic anisotropy of Fe/Pt(001) and Pt/Fe/Pt(001) using a first-principles approach, *Phys. Rev. B*, **77**, 054413 (2008).
79. C. Clavero, J. R. Skuza, Y. Choi, D. Haskel, J. M. García-Martín, A. Cebollada, and R. A. Lukaszew, Control of the perpendicular magnetic anisotropy of FePd films via Pd capping deposition, *Appl. Phys. Lett.*, **92**, 162502 (2008).
80. T. Yang, E. Ahmad, and T. Suzuki, FePt–Ag nanocomposite film with perpendicular magnetic anisotropoy, *J. Appl. Phys.*, **91**, 6860–6862 (2002).
81. C. P. Luo, S. H. Liou, and D. J. Sellmyer, FePt:SiO_2 granular thin film for high density magnetic recording, *J. Appl. Phys.*, **87**, 6941–6943 (2000).
82. C. P. Luo, S. H. Liou, L. Gao, Y. Liu, and D. J. Sellmyer, Nanostructured FePt:B_2O_3 thin films with perpendicular magnetic anisotropy, *Appl. Phys. Lett.*, **77**, 2225–2227 (2000).
83. J. I. Martín, J. Nogués, K. Liu, J. L. Vicent, and I. K. Schuller, Ordered magnetic nanostructures: fabrication and properties, *J. Magn. Magn. Mater.*, **256**, 449–501 (2003).

84. T. Bublat and D. Goll, Large-area hard magnetic $L1_0$-FePt nanopatterns by nanoimprint lithography, *Nanotechnology*, **22**, 315301 (2011).
85. R. L. White, R. M. H. Newt, and R. F. W. Pease, Patterned media: a viable route to 50 Gbit/in^2 and up for magnetic recording?, *IEEE Trans. Magn.*, **33**, 990–995 (1997).
86. G. F. Hughes, Patterned media, in *The Physics of Ultra-High-Density Magnetic Recording*, 205–229, M. L. Plumer, J. van Ek, and D. Weller, Eds. Berlin: Springer-Verlag, 2001.
87. D. Weller and M. F. Doerner, Extremely high-density longitudinal magnetic recording media, *Annu. Rev. Mater. Sci.*, **30**, 611–644 (2000).
88. B. D. Terris and T. Thomson, Nanofabricated and self-assembled magnetic structures as data storage media, *J. Phys. D: Appl. Phys.*, **38**, R199–R222 (2005).
89. Q. Y. Xu, Y. Kageyama, and T. Suzuki, Ion-beam-induced chemical-vapor deposition of FePt and CoPt particles, *J. Appl. Phys.*, **97**, 10K308 (2005).
90. A. Lyberatos and K. Yu Guslienko, Thermal stability of the magnetization following thermomagnetic writing in perpendicular media, *J. Appl. Phys.*, **94**, 1119–1129 (2003).
91. C. T. Boone, J. A. Katine, E. E. Marinero, S. Pisana, and B. D. Terris, Demonstration of microwave assisted magnetic reversal in perpendicular media, *J. Appl. Phys.*, **111**, 07B907 (2012).
92. D. Suess, T. Schrefl, S. Fähler, M. Kirschner, G. Hrkac, F. Dorfbauer, and J. Fidler, Exchange spring media for perpendicular recording, *Appl. Phys. Lett.*, **87**, 012504 (2005).
93. A. Yu Dobin and H. J. Richter, Domain wall assisted magnetic recording, *Appl. Phys. Lett.*, **89**, 062512 (2006).
94. H. Wang, M. Tofizur Rahman, H. Zhao, Y. Isowaki, Y. Kamata, A. Kikitsu, and J.-P. Wang, Fabrication of FePt type exchange coupled composite bit patterned media by block copolymer lithography, *J. Appl. Phys.*, **109**, 07B754 (2011).

Chapter 4

Relaxation in Magnetic Materials for Spintronics

Claudia K. A. Mewes and Tim Mewes

Center for Materials for Information Technology,
Department of Physics and Astronomy, University of Alabama,
Tuscaloosa, AL 35487, USA
cmewes@mint.ua.edu, tmewes@mint.ua.edu

In recent years, extensive research aimed at improving the understanding of the ultrafast magnetization dynamics has been carried out. Novel phenomena, for example the influence of the lateral confinement in small structures on the spin dynamics [1–2] and the spin dynamics due to spin-polarized currents [4–8] make this a very interesting area for fundamental research. The magnetization dynamics are, on the other hand, also of crucial importance for the application of magnetic materials in current and future spintronic devices like hard drive read heads, magnetic random access memories (MRAMs), and spin torque oscillators. The relaxation properties in these materials and devices are particularly important in view of the power consumption of spintronic devices.

Handbook of Nanomagnetism: Applications and Tools
Edited by Rosa A. Lukaszew

ISBN 978-981-4613-04-0 (Hardcover), 978-981-4613-05-7 (eBook)
www.panstanford.com

The magnetization dynamics can be described by the Landau–Lifshitz–Gilbert equation of motion [9, 10]:

$$\frac{d\vec{M}}{dt} = -\gamma\vec{M} \times \vec{H}_{eff} + \frac{1}{M_S}\vec{M} \times \alpha\frac{d\vec{M}}{dt} \tag{4.1}$$

where $\vec{M}$ is the magnetization vector, M_S is the saturation magnetization, γ is the gyromagnetic ratio and α is the damping parameter. The effective field $\vec{H}_{eff}$ entering the Landau–Lifshitz–Gilbert equation of motion is given by the negative derivative of the Gibbs free energy density with respect to the components of the magnetization vector:

$$\vec{H}_{eff} = -\frac{\partial E}{\partial \vec{M}} \tag{4.2}$$

The effective field is the sum of all fields acting on the magnetization, including the external magnetic field $\vec{H}$, anisotropy fields, exchange fields, and dipolar fields. It can be shown [9, 10] that the Landau–Lifshitz–Gilbert equation is mathematically equivalent to the older Landau–Lifshitz equation of motion if the gyromagnetic ratio and damping parameters are transformed properly [11]. In addition to the dimensionless damping parameter α, the relaxation rate $\lambda = G = \gamma\alpha M_S$ is also used in the literature.

Over the years the Landau–Lifshitz–Gilbert equation of motion has been applied to a wide range of magnetization dynamics using either a macrospin model or full micromagnetic simulations. This includes, for example, domain wall motion [12, 13], vortex dynamics [14], ultrafast magnetization reversal [15], and spin transfer torque [16]. Similarly there are a number of experimental techniques that have been developed to characterize the magnetization dynamics and can shed light on the relaxation of the magnetization. Among those are ferromagnetic resonance (FMR) [26] and Brillouin light scattering (BLS) [17], which probe the small-precession-angle dynamics of the magnetization. On the other hand time-resolved techniques, including inductive measurements [18], magneto-optical Kerr effect magnetometry [19], and X-ray magnetic circular dichroism [20], enable the investigation of switching events and the associated large-angle dynamics that often lead to nonlinear damping [21–23]. In the following section the focus will be on small-angle dynamics and their relaxation of magnetization.

In magnetic resonance experiments, one applies a static external magnetic field $\vec{\mathrm{H}}$ and an external radiofrequency magnetic field $\vec{h}_{\mathrm{rf}}$. In this case, the equation of motion can be utilized to predict the condition to achieve FMR [24–27]. For small precession angles of magnetization around its equilibrium position, the condition for FMR is known as the Smit–Beljers equation [25, 28, 29]:

$$\left(\frac{\omega}{\gamma}\right)^2 = \frac{1}{M_s^2 \sin^2\theta}\left[\frac{\partial^2 E}{\partial\theta^2}\frac{\partial^2 E}{\partial\varphi^2} - \left(\frac{\partial^2 E}{\partial\varphi\partial\theta}\right)^2\right] \tag{4.3}$$

where ω is the resonance frequency and the second derivatives of the Gibbs free energy density E are taken at the points of equilibrium, that is, $\left.\frac{\partial E}{\partial\varphi}\right|_{\varphi_0} = \left.\frac{\partial E}{\partial\theta}\right|_{\theta_0} = 0$.

In FMR experiments, the frequency is typically held constant, while the magnetic field is swept through resonance [26]. However, using vector network analyzer FMR it is also possible to conduct experiments at a fixed field while sweeping the frequency [30, 31]. The damping term in the Landau–Lifshitz–Gilbert equation of motion leads to a finite width of the resonance absorption. The width of the resonance in field-swept experiments is characterized by the FMR peak-to-peak linewidth ΔH_{PP}, that is, the field between the two inflection points of the resonance curve. On the basis of Suhl's formula for the linewidth [32]

$$\Delta H_{\mathrm{PP}} = \frac{1}{\sqrt{3}}\frac{\alpha}{|d\omega/dH|}\frac{\gamma}{\mathrm{M_S}}\left[\frac{\partial^2 E}{\partial\theta^2} + \frac{1}{\sin^2\theta}\frac{\partial^2 E}{\partial\varphi^2}\right] \tag{4.4}$$

one can obtain an approximate expression that is proportional to the resonance frequency [33–37]:

$$\Delta H_{\mathrm{PP}} \approx \frac{2}{\sqrt{3}}\frac{\alpha}{\gamma}\frac{\omega}{\cos\beta} \tag{4.5}$$

where β is the angle between the magnetization and the external magnetic field. Due to the $\cos\beta$ term in the denominator the FMR linewidth is enhanced whenever the magnetization is not aligned with the external magnetic field; accordingly this effect is called field dragging [37]. On the other hand, if the external magnetic field is aligned with an easy or hard direction of the magnetization, the magnetization is parallel to the field direction for sufficiently large fields [38], that is, $\beta = 0$, and the dragging contribution vanishes, resulting in

$$\Delta H_{\mathrm{PP}} \approx \frac{2}{\sqrt{3}} \frac{\alpha}{\gamma} \omega. \tag{4.6}$$

Unfortunately this relationship is often misinterpreted to indicate that, given a measured linewidth at one particular frequency, all one has to do to obtain the damping parameter is to solve Eq. 4.6 for α. From the discussion of the different damping mechanisms below it will become clear that instead, the experimentalist should first verify that the linewidth is indeed a linear function of the microwave frequency.

One common modification of the above equation is to include a zero-frequency offset ΔH_0 [33, 34, 39]:

$$\Delta H_{\mathrm{PP}} = \Delta H_0 + \frac{2}{\sqrt{3}} \frac{\alpha}{\gamma} \omega \tag{4.7}$$

The zero-frequency offset characterizes contributions to the resonance linewidth from sample inhomogeneities that result in a spread of resonance frequencies throughout the sample [40].

4.1 Damping Mechanisms

Originally, the damping term in Eq. 4.1 was introduced as a phenomenological description of the relaxation of the magnetization [9]. While this phenomenological description has been very successful in describing the magnetization dynamics, the details of its physical origin are still being actively investigated. One difficulty that arises when trying to compare experimental results with theoretical calculations is that there are a number of mechanisms that lead to a term proportional to the vector product of the magnetization and its time derivative, that is, have the form of the damping term in the Landau–Lifshitz–Gilbert equation of motion. Generally one can distinguish between contributions to the magnetic relaxation that are intrinsic and extrinsic to the material system. One considers a contribution to the relaxation intrinsic to the material when it is unavoidable under the given conditions [27]. Conversely, contributions that are avoidable, like those caused by inhomogeneities or other geometrical peculiarities of the sample should be called extrinsic [27].

4.2 Magnon–Phonon Interaction

At finite temperatures thermally excited phonons are unavoidable and thus can lead to a contribution to the damping of a material. However, as pointed out by Heinrich [27], damping caused by direct magnon–phonon coupling, as calculated by Suhl [41], is significantly smaller than experimentally observed values. While the first-order contribution of the magnon–phonon coupling to the relaxation is of the form of the damping term in the Landau–Lifshitz–Gilbert equation of motion, it is noteworthy that Suhl also derived terms that do not have the Landau–Lifshitz–Gilbert form [41]. In an earlier work by Kobayashi et al. [42], a calculation based on the solution of the coupled magnetoelastic equations of motion has shown that an appreciable effect can only be observed when the elastic wave forms a standing wave across the film thickness near the ferromagnetic resonance (FMR) frequency. Thus, in summary, magnon–phonon coupling does not appear to contribute significantly to the damping. However, more experimental work [43, 44] is needed to further clarify the magnon–phonon interaction and its implications for the relaxation of magnetization.

4.3 Eddy Current Damping

In a conductor, time-varying magnetic fields, and thus a time-varying magnetization, will induce eddy currents [45, 46]. The problem of eddy currents for a time-varying, nonuniform magnetization generally requires one to solve the Landau–Lifshitz–Gilbert equation of motion and the Maxwell equations simultaneously [47–51]. However, for ultrathin films, which are at the focus of most current research, the eddy current contribution to magnetization relaxation again has the form of the damping term in the Landau–Lifshitz–Gilbert equation of motion. Furthermore for films thinner than the skin depth, one expects the eddy current contribution to the damping to scale as t_F^2, where t_F is the film thickness, while also being proportional to the electrical conductivity of the material [27, 46, 52]. For films thicker than the skin depth, eddy currents result in a spatial gradient in the dynamic component of the magnetization and the exchange interaction must be taken into account. For this reason, this contribution to the relaxation is also referred to as the exchange conductivity mechanism [27, 45]. It is noteworthy that the exchange conductivity mechanism also leads to an asymmetry in the resonance lineshape [53–55].

In summary, one expects the eddy current contribution to damping to be small in ultrathin films [27, 52, 56] and the quadratic dependence on the film thickness offers a possibility to experimentally detect its presence.

4.4 Spin–Orbit Relaxation

Spin–orbit relaxation can be described using what is known as Kamberský's torque correlation model [57, 58]. Recent work by Gilmore [59, 60] and Kamberský [61] based on first-principles calculations has shown that this mechanism is the dominant relaxation process in metallic ferromagnets. Within this model, the relaxation rate is determined by the decay of the magnons into electron–hole pairs and the scattering of those electron–hole pairs with phonons or lattice defects. When calculating the relaxation rate, the electron–hole pair creation is treated explicitly, while the subsequent scattering with phonons or lattice defects is typically treated using a phenomenological lifetime τ for the states [59–62]. For the electron–hole pair creation, one distinguishes between intraband transitions that occur within a single band and interband transitions that occur between two different bands. Both contributions can also be formulated using an effective field, and the electronic energy of the system is given by [60]

$$E = \sum_{n,k} \rho_{nk} \epsilon_{nk} \tag{4.8}$$

that is, a summation over the single-electron energies ϵ_{nk} weighted by the state occupancies ρ_{nk}, where n indicates the band index and

k the wave vector. On the basis of Eq. 4.2 this gives the following contributions to the effective field:

$$\vec{H}_{\text{eff}} = -\sum_{n,k}\left[\rho_{nk}\frac{\partial\epsilon_{nk}}{\partial\vec{M}} + \frac{\partial\rho_{nk}}{\partial\vec{M}}\epsilon_{nk}\right] \tag{4.9}$$

The first term is the contribution that arises from changes of the energy of each state as the magnetization precesses and is also called the breathing Fermi surface model [60, 63, 64]. As demonstrated by Gilmore et al. [60] this term corresponds to the intraband transitions in Kamberský's torque correlation model. The intraband contribution to the relaxation increases linearly with the lifetime τ and thus is also called a conductivity-like term. The second term in the above equation corresponds to the interband transitions and can be visualized as the "bubbling" of electrons on the Fermi surface [60]. The interband contribution to the relaxation is roughly inversely proportional to the lifetime and is therefore called a resistivity-like term. Correspondingly the intraband contribution decreases with increasing temperature, whereas the interband contribution increases with increasing temperature. Therefore, the interband contribution will dominate the relaxation at sufficiently high temperatures, whereas at low temperatures, the intraband contribution will become dominant. This behavior can lead to a minimum in the spin–orbit damping as a function of temperature, which resembles the experimentally observed behavior for Ni [65, 66]. Kamberský's torque correlation model also explains the low relaxation observed in half metals [67–69], as the half metallicity implies the absence or reduction of contributions involving the spin channel with a gap at the Fermi energy [62] (compare Fig. 4.1).

Recently it has been pointed out that Kamberský's torque correlation model also leads to an anisotropic damping of the magnetization [63, 64, 71–73]. The anisotropy of the damping can be in the form of a rotational anisotropy, for which the scalar damping parameter α has to be replaced by a damping matrix $\underline{\underline{\alpha}}$ that takes into account that the damping is different for different orientations of $\frac{d\vec{M}}{dt}$. It can also lead to an orientational anisotropy, for which the damping depends on the current orientation of the magnetization $\vec{M}$. Furthermore, the damping can also become nonlocal such that the damping matrix depends on the orientation of

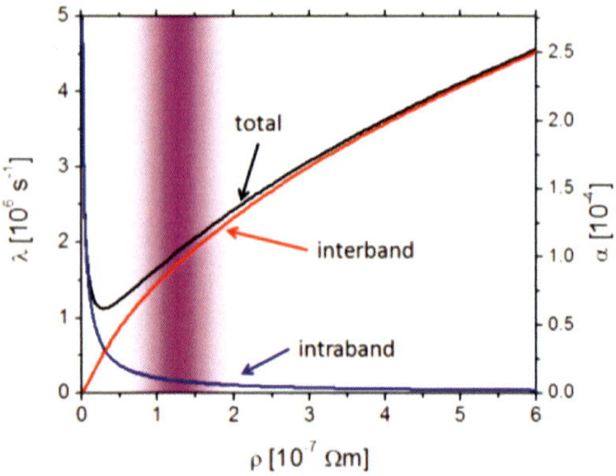

Figure 4.1 Damping rate λ and damping parameter α caused by spin-orbit relaxation for the half-metal Co_2MnSi as a function of resistivity. The resistivity-like interband contribution and the conductivity-like intraband contribution lead to a minimum of the total damping parameter [62]. The region highlighted in purple represents resistivity values reported in Ref. [70].

all magnetic moments in the system [71, 74]. Gilmore and Stiles [72] point out that the interband contributions to the damping exhibit rotational and orientational anisotropy only for small scattering rates (low temperatures) and become more isotropic for increasing scattering rates. The intraband contribution, on the other hand, is predicted to maintain both rotational and orientational anisotropy for all scattering rates. However, it will be quite challenging to experimentally observe these anisotropies of the damping. First of all, by replacing the scalar damping parameter in the Landau–Lifshitz–Gilbert equation of motion with a damping matrix, the usual approach to analyze FMR data using Eq. 4.5 is no longer justified. Instead one has to resort to numerically solving the Landau–Lifshitz–Gilbert equation and compare the numerical solution with the experimental data [71]. Furthermore, as pointed out by Woltersdorf and Heinrich [75], an apparent angular dependence of damping can also result from lattice defects and two-magnon scattering. In this context it is also noteworthy that, besides anisotropic damping, Kamberský's torque correlation model also leads to an inertial damping term that does not have the Landau–

Lifshitz–Gilbert form [73] but instead contains the second time derivative of the magnetization $\frac{d^2\vec{M}}{dt^2}$, similar to Suhl's earlier results on magnon–phonon damping [41].

In summary the spin–orbit relaxation is currently considered to be the dominant relaxation process in metallic ferromagnets. Its close relationship with the conductivity and resistivity of the material suggests that temperature-dependent measurements of the damping will remain important to further test Kamberský's torque correlation model. Recent theoretical investigations have revealed that the torque correlation model leads to anisotropic damping and an inertial damping term. This should come as a welcome challenge for the experimentalist to try to devise experiments that can test these predictions with sufficient accuracy.

4.5 Spin Pumping

When investigating the relaxation of magnetization in thin films, the samples typically have the ferromagnetic film of interest sandwiched between other layers. If the adjacent layers are conducting, their presence can give rise to an additional contribution to the relaxation of the magnetization known as spin pumping [76–78]. A precessing magnetization in the ferromagnet leads to a spin current $\vec{I}_S^{\text{pump}}$ that is pumped out of the ferromagnet into the adjacent layers [76]:

$$\vec{I}_S^{\text{pump}} = \frac{\hbar}{4\pi}\left(A_r \hat{m} \times \frac{d\hat{m}}{dt} - A_i \frac{d\hat{m}}{dt}\right) \tag{4.10}$$

where $A = A_r + iA_i$ is the complex-valued spin pumping conductance and $\hat{m}$ is the magnetization direction. The imaginary part of the spin-pumping conductance will lead to a shift of the resonance position in magnetic resonance experiments. However, first-principles calculations indicate that A_i is very small for Cu/Co and Fe/Cr interfaces [79] and can likely be disregarded in many systems [76]. As can be seen in Eq. 4.10, the term associated with the real part of the spin-pumping conductance has the same form as the damping term in the Landau–Lifshitz–Gilbert equation of motion and will thus lead to an increased damping of the magnetization. The spin current $\vec{I}_S^{\text{pump}}$ will lead to a spin accumulation in the adjacent normal metal layers, which in turn creates a backflow of spin current $\vec{I}_S^{\text{back}}$, but can also relax by spin-flip scattering. For a perfect spin sink, that is, a material with an infinite spin-flip rate, one has $\vec{I}_S^{\text{back}} = 0$, which gives the maximum increase in damping due to spin pumping. For a poor spin sink on the other hand one has $\vec{I}_S^{\text{back}} \approx \vec{I}_S^{\text{pump}}$ and there will be very little effect of the adjacent layers on the damping in the ferromagnet. Due to the nonlocal nature of the damping contribution caused by spin pumping, one expects this contribution to scale with the inverse thickness of the ferromagnet [77] for a reasonable good spin sink, which is what has been observed experimentally [80–83]. Using Onsager's reciprocity relations it has been shown that spin pumping is the reciprocal phenomenon to spin transfer torque [84–86].

Spin pumping can become a significant contribution to the relaxation of the magnetization in thin films adjacent to normal metals and it scales as the inverse thickness of the ferromagnet. Furthermore, spin pumping is an important ingredient for devices based on spin currents, as the FMR creates a pure spin source or "spin battery" [87].

4.6 Two-Magnon Scattering

For ferromagnetic insulators, two-magnon scattering was recognized early on as an important contribution to the relaxation of the magnetization [88–90]. Later, it has been shown that two-magnon scattering can also contribute significantly to the relaxation in thin metallic ferromagnetic films, provided the magnetization lies in the plane of the film [91–94]. For FMR experiments, two-magnon scattering describes the process through which the uniform precession mode with wave vector $\mathrm{k} = 0$ scatters into degenerate (short-wavelength) spin waves with a nonzero wave vector parallel to the film surface $\mathrm{k}_{\parallel} \neq 0$, thereby contributing to the relaxation of the uniform precession mode. Therefore, for two-magnon scattering to be operative, degenerate spin waves with a finite wave vector are required as well as inhomogeneities that act as scattering centers. It is instructive to consider the spin-wave dispersion for a thin film with (a) the field applied in the film plane and (b) the field applied perpendicular to the film plane; a more general and detailed discussion can be found in Ref. [93].

In the case when the field is applied in the film plane the spin-wave dispersion for a thin film with an exchange stiffness $D = 2A/M_S$ can be expressed as follows [93]:

$$\left(\frac{\omega\left(\vec{k}_{\|}\right)}{\gamma}\right)^2 = \left(\frac{\omega_{\mathrm{FMR}}}{\gamma}\right)^2 - H_1 \pi M_S t_F \cdot k_{\|} + H_2 D \cdot k_{\|}^2 \qquad (4.11)$$

where the frequency of the uniform precession is given by

$$\left(\frac{\omega_{\mathrm{FMR}}}{\gamma}\right)^2 = H \cdot (H + 4\pi M_{\mathrm{eff}}) \qquad (4.12)$$

where the effective magnetization $4\pi M_{\mathrm{eff}}$ also includes a possible out-of-plane anisotropy H_a, that is, $4\pi M_{\mathrm{eff}} = 4\pi M_s + H_a$. Furthermore $H_1 = 2(H - (H + 4\pi M_{\mathrm{eff}}) \sin^2 \phi_k)$ with ϕ_k indicating the angle between the spin wave vector $\vec{k}_{\|}$ and the direction of the magnetization and $H_2 = 2H + 4\pi M_{\mathrm{eff}} > 0$. Therefore, when the external magnetic field is applied in the plane of the film, there are always spin waves that are degenerate with the uniform precession. The only requirement is that the propagation directions of the spin waves fulfill

$$\sin^2 \phi_k < \frac{H}{H + 4\pi M_{\mathrm{eff}}} \qquad (4.13)$$

The situation changes, however when the field is applied perpendicular to the film plane, and the dispersion relation for $H > 4\pi M_{\mathrm{eff}}$ is given by [93]

$$\left(\frac{\omega\left(\vec{k}_{\|}\right)}{\gamma}\right)^2 = \left(\frac{\omega_{\mathrm{FMR}}}{\gamma}\right)^2 + H_3 (\pi M_S t_F \cdot k_{\|} + D \cdot k_{\|}^2) \qquad (4.14)$$

where the frequency of the uniform precession is given by

$$\frac{\omega_{\mathrm{FMR}}}{\gamma} = H - 4\pi M_{\mathrm{eff}} \qquad (4.15)$$

and $H_3 = 2\,(H - 4\pi M_{\mathrm{eff}}) > 0$. In this case the coefficients for the terms linear and quadratic in $k_{\|}$ have the same sign and therefore no spin waves degenerate with the uniform precession exist. Thus there is no two-magnon contribution to the magnetization relaxation

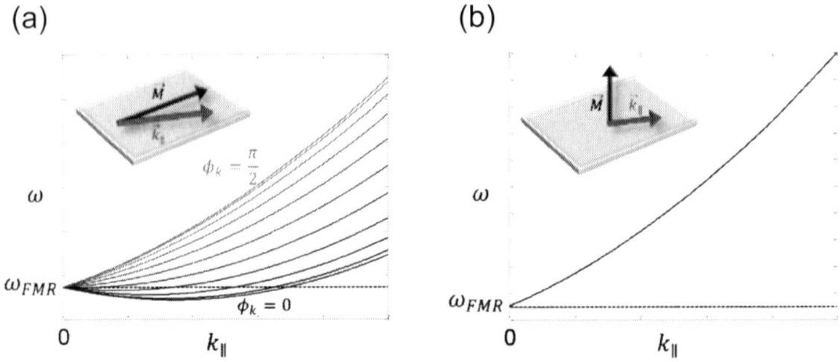

Figure 4.2 Spin-wave dispersion of a thin film with the external magnetic field and magnetization (a) in the film plane and (b) perpendicular to the film plane. The lowest branch in (a) shows the dispersion for spin waves propagating along the direction of the magnetization whereas the highest branch is for spin waves propagating perpendicular to the magnetization. In both graphs the uniform precession is shown as a dashed line.

when the external magnetic field is applied perpendicular to the film (see Fig. 4.2).

To conserve momentum two-magnon scattering requires the presence of defects, which naturally will occur at surfaces and interfaces of the film. These structural defects then couple the uniform precession to the degenerate spin waves through their associated perturbation of the Zeeman field, the dipole field, or the anisotropy field [93, 95]. Therefore the calculation of the contribution of the two-magnon scattering to the damping requires a quantitative model for the interfacial/surface roughness of the ferromagnetic film. However, using the assumptions made by Arias and Mills, the following expression can be obtained for two-magnon contribution $\Delta H^{(2)}$ to the ferromagnetic linewidth [93, 94, 96]:

$$\Delta H^{(2)} = \Gamma(\omega) \arcsin \sqrt{\frac{\sqrt{\left(\frac{\omega_0}{2}\right)^2 + \omega_{\mathrm{FMR}}^2} - \frac{\omega_0}{2}}{\sqrt{\left(\frac{\omega_0}{2}\right)^2 + \omega_{\mathrm{FMR}}^2} + \frac{\omega_0}{2}}} \tag{4.16}$$

with $\omega_0 = \gamma 4\pi \mathrm{M}_{\mathrm{eff}}$. The prefactor $\Gamma(\omega)$ indicates the strength of the two-magnon scattering and, as described above, depends on the details of the interfacial roughness. If one assumes the perturbations to be strictly interfacial, then the scattering strength scales as $1/t_{\mathrm{F}}^2$, that is, like the inverse film thickness squared [92, 93, 97]. If one sets

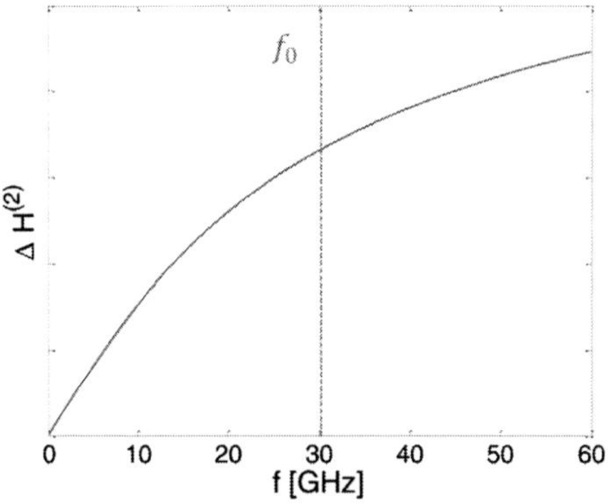

Figure 4.3 Two-magnon contribution to the linewidth based on Eq. 4.16 for material parameters corresponding to permalloy The frequency $f_0 = \omega_0/2\pi$ is comparable to microwave frequencies used in typical ferromagnetic resonance experiments.

aside the frequency dependence of the scattering strength, which according to Ref. [93], can be assumed to be weak, one obtains the dependence shown in Fig. 4.3. Nonlinear frequency dependencies of the FMR linewidth consistent with Eq. 4.16 have since been observed in numerous experiments [36, 75, 98–101].

However, depending on the experimentally available frequency range and the magnitude of the two-magnon contribution compared to other linewidth contributions, the deviation from linear frequency dependence may be difficult to establish. Because the two-magnon scattering strength is determined by interfacial defects, it will reflect the in-plane angular symmetry of the defect structures [36, 75, 93, 98]. This has recently been demonstrated using artificially introduced periodic defects [102]. However, due to the periodic nature of the defects in this study, this leads to a periodic enhancement of the linewidth rather than the monotonic frequency dependence predicted by Eq. 4.16. In experiments on exchange-biased films a unidirectional contribution to the damping has been attributed to two-magnon scattering on the basis of its dependence on the inverse square of the film thickness [103]. In this case the angular dependence of the relaxation reflects the unidirectional symmetry of the exchange bias field, which acts as the perturbation field at the interface.

Two-magnon scattering can contribute significantly to the relaxation in ultrathin films magnetized in the film plane but is absent in films magnetized perpendicular to the film plane. The two-magnon scattering strength depends on the interfacial roughness of the film and is expected to scale as the inverse film thickness squared. In FMR experiments, the two-magnon scattering causes a nonlinear frequency dependence of the linewidth, which distinguishes this contribution to the relaxation from others that have the Landau–Lifshitz–Gilbert form.

4.7 Slow-Relaxing Impurities

The slow relaxer mechanism was originally introduced by van Vleck and Orbach [104] to explain experimental observations regarding the relaxation in rare earth–doped yttrium iron garnet (YIG) [105–107]. However, since then it has also been applied to $(Co_2Mn)_{1-x}Ge_x$ thin films [108], rare earth impurities in $Ni_{80}Fe_{20}$ [109] and in a modified form to a number of exchange bias systems [110–113]. The basic idea of the slow relaxer model is that the splitting of the lowest-energy levels of the rare earth ion is modulated by the precessing magnetization. The population of the two energy levels will tend to reestablish thermal equilibrium, leading to transitions between them. These thermal transitions, in turn, lead to a fluctuating field acting on the precessing magnetization, thereby contributing to the relaxation of the magnetization. This contribution is of the Landau–Lifshitz–Gilbert form but the damping parameter α_{slow} is frequency dependent [108, 109, 114, 115]:

$$\alpha_{\text{slow}}(\omega) = C \cdot F \frac{\tau}{1+(\omega\tau)^2} \tag{4.17}$$

with

$$C(T) = \frac{|\gamma| N}{8 M_S k_B T}\left(\left(\frac{\partial E_{\text{slow}}}{\partial \phi}\right)^2 + \left(\frac{\partial E_{\text{slow}}}{\partial \theta}\right)^2\right) \tag{4.18}$$

and

$$F(T) = \sec h^2\left(\frac{E_{\text{slow}}}{k_B T}\right) \tag{4.19}$$

τ is the relaxation time, N is the concentration, and E_{slow} is the energy splitting of the slow-relaxing impurity. From Eq. 4.18 it follows that for an isotropic energy splitting of the slow-relaxing impurity the damping parameter α_{slow} is zero. It should be noted that the relaxation time τ itself depends strongly on the temperature [116–118]. The frequency-dependent damping parameter for slow-relaxing impurities is maximal for zero frequency, but the contribution of the slow relaxer to the FMR linewidth has a maximum at $\omega = 1/\tau$ due to the additional factor ω (see Eq. 4.6). However, the above model has recently been extended to allow for coherence in the slow-relaxing impurity, leading to a peak in the damping parameter at a nonzero frequency [108]. For short relaxation times the maximum in the FMR linewidth may be at a frequency outside the experimentally available frequency range (see $T = 300$ K data in Fig. 4.4). In the limit $\omega \ll 1/\tau$ the damping contribution of the slow relaxer in Eq. 4.17 becomes independent of the frequency [109]. With decreasing temperature the relaxation time increases and therefore the linewidth peak will shift to lower frequencies. Consequently a measurement of the linewidth at a fixed frequency will also exhibit a maximum as a function of temperature (see Fig. 4.4).

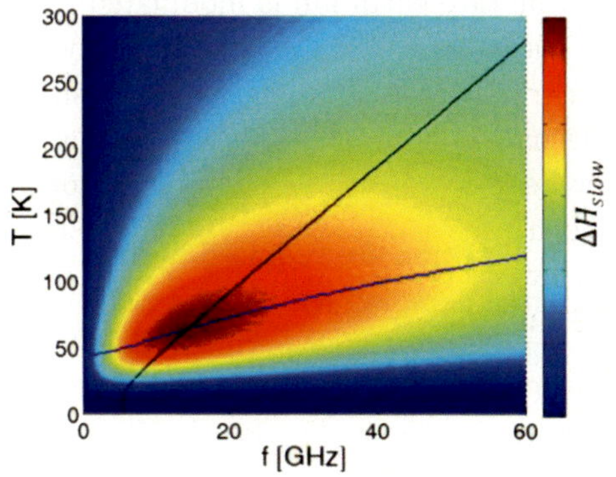

Figure 4.4 Contribution of the slow relaxer to the linewidth as a function of frequency and temperature. Here an energy splitting of $E_{\text{slow}} = 50\ k_B$ and a relaxation time at $T = 0$ of $\tau_0 = 30$ ps was assumed. Further a direct process spin–lattice relaxation was assumed [116]. The maximum in the frequency dependence of the linewidth for constant temperatures is shown as a black line. The maximum in the temperature dependence of the linewidth for constant frequencies is shown as a blue line.

The slow-relaxing impurity model has been successfully applied to a number of thin metallic films. It predicts a frequency-dependent contribution to the damping proportional to the impurity concentration. Combining frequency and temperature-dependent measurements should provide further insights into the nature of the slow relaxer and its relaxation mechanism.

4.8 Conclusions and Outlook

There are a number of relaxation mechanisms for the magnetization in thin films that have the form of the damping term used in the Landau–Lifshitz–Gilbert equation of motion. Because of this it can be difficult to experimentally distinguish the different contributions to the relaxation. FMR experiments are a valuable tool to investigate the magnetization dynamics in the small-precession-angle limit. The recent progress in performing FMR experiments over a broad frequency range enables experimental investigation of the predicted linear relationship between ferromagnetic linewidth and microwave frequency (Eq. 4.7) with unprecedented accuracy, which is the first step in determining the importance of the different contributions to the relaxation. Additional experiments varying the film thickness, adjacent layers, the direction of the applied field, and the temperature are required to further pinpoint the mechanisms responsible for magnetization relaxation. The theoretical prediction that modifications of the damping term of the Landau–Lifshitz–Gilbert equation of motion are required to account for anisotropic damping and inertial damping will require careful analysis in order to avoid misinterpretation of experimental data. Thus, although introduced more than 50 years ago, the damping term in the equation of motion for magnetization still remains a very active and important area of research which will influence the future development of the field of spintronics.

Acknowledgments

We would like to thank all our colleagues for stimulating discussions and suggestions that helped us in preparing this chapter. In particular we would like to thank Dr. W. H. Butler, Mr. B. Khodadadi, Ms. J B. Mohammadi, Mr. S. Paul, Dr. S. Schäfer and Dr. M. Williams. We would also like to thank Dr. H. T. Nembach, Dr. T. J. Silva and Dr. G. Woltersdorf for the many fruitful discussions we had with them and in particular for their invaluable advice on the mechanism of slow-relaxing impurities. Further we would like to thank our collaborators Dr. M. J. Carey, Dr. J. R. Childress, Dr. S. Maat, Dr. B. York, Dr. C. Kaiser, Dr. Q. Leng and Dr. M. Pakala. We would like to thank Dr. R. L. Stamps for all the inspiring discussion we had with him over the years and in particular for his invaluable insights regarding two-magnon scattering. We would also like to thank Dr. R. D. McMichael and Dr. M. Stiles for the stimulating discussions during our stay at NIST in Gaithersburg.

We also would like to acknowledge support by the National Science Foundation, in particular through grants DMR 0804243 CAREER 0952929, and CAREER 1452670.

References

1. W. K. Hiebert, A. Stankiewicz, M. R. Freeman, Direct observation of magnetic relaxation in a small permalloy disk by time-resolved scanning Kerr microscopy, *Phys. Rev. Lett.,* **79**, 1134 (1997).
2. J. Jorzick, S. O. Demokritov, B. Hillebrands, B. Bartenlian, C. Chappert, D. Decanini, F. Rousseaux, E. Cambril, Spin-wave quantization and dynamic coupling in micron-size circular magnetic dots, *Appl. Phys. Lett.,* **75**, 3859 (1999).
3. J. P. Park, P. Eames, D. M. Engebretson, J. Berezovsky, P. A. Crowell, Imaging of spin dynamics in closure domain and vortex structures, *Phys. Rev. B,* **67**, 020403 (2003).
4. S. I. Kiselev, J. C. Sankey, I. N. Krivorotov, N. C. Emley, R. J. Schoelkopf, R. A. Buhrman, D. C. Ralph, Microwave oscillations of a nanomagnet driven by a spin-polarized current, *Nature,* **425**, 380–383 (2003).
5. W. H. Rippard, M. R. Pufall, S. Kaka, S. E. Russek, T. J. Silva, Direct-current induced dynamics in $Co_{90}Fe_{10}/Ni_{80}Fe_{20}$ point contacts, *Phys. Rev. Lett.,* **92**, 027201 (2004).
6. S. Kaka, M. R. Pufall, W. H. Rippard, T. J. Silva, S. E. Russek, J. A. Katine, Mutual phase-locking of microwave spin torque nano-oscillators, *Nature,* **437**, 389–392 (2005).

7. I. N. Krivorotov, N. C. Emley, J. C. Sankey, S. I. Kiselev, D. C. Ralph, R. A. Buhrman, Time-domain measurements of nanomagnet dynamics driven by spin-transfer torques, *Science*, **307**, 5707 (2005).
8. A. A. Tulapurkar, Y. Suzuki, A. Fukushima, H. Kubota, H. Maehara, K. Tsunekaw, D. D. Djayaprawira, N. Watanabe, S. Yuasa, Spin-torque diode effect in magnetic tunnel junctions, *Nature*, **438**, 339–342 (2005).
9. T. L. Gilbert, A. Lagrangian formulation of the gyromagnetic equation of the magnetization fields (abstract only), *Phys. Rev.*, **100**, 1243 (1955); T. L. Gilbert, A phenomenological theory of damping in ferromagnetic materials, *IEEE Trans. Magn.*, **40**, 3443 (2004).
10. A. Aharoni, *Introduction to the Theory of Ferromagnetism*, Oxford University Press, 2000.
11. See, for example, M. D. Stiles, W. M. Saslow, M. J. Donahue, A. Zangwill, Reply to "comment on 'adiabatic domain wall motion and Landau-Lifshitz damping'," *Phys. Rev. B*, **78**, 216402 (2008); G. Bertotti, I. D. Mayergoyz, C. Serpico, Identification of the damping coefficient in Landau–Lifshitz equation, *Phys. B*, **306**, 102 (2001).
12. J. C. Slonczewski, Theory of domain-wall motion in magnetic films and platelets, *J. Appl. Phys.*, **44**, 1759 (1973).
13. V. V. Volkov, V. A. Bokov, Domain wall dynamics in ferromagnets, *Phys. Solid State*, **50**, 199 (2008) and references therein.
14. R. Antos, Y. Otani, J. Shibata, Magnetic vortex dynamics, *J. Phys. Soc. Jpn.*, **77**, 031004 (2008) and references therein.
15. A. Kirilyuk, A. V. Kimel, T. Rasing, Ultrafast optical manipulation of magnetic order, *Rev. Mod. Phys.*, **83**, 2731 (2010) and references therein.
16. D. V. Berkov, J. Miltat, Spin-torque driven magnetization dynamics: micromagnetic modeling, *J. Magn. Magn. Mater.*, **320**, 1238 (2008) and references therein.
17. B. Hillebrands, Brillouin light scattering spectroscopy, in *Modern Techniques for Characterizing Magnetic Materials*, Springer, 2005, and references therein.
18. T. J. Silva, C. S. Lee, T. M. Crawford, C. T. Rogers, Inductive measurement of ultrafast magnetization dynamics in thin-film permalloy, *J. Appl. Phys.*, **85**, 7849 (1999).
19. J. Fassbender, Magnetization dynamics investigated by time-resolved Kerr effect magnetometry, in *Spin Dynamics in Confined Magnetic Structures II*, Springer, 2003.

20. S. Pizzini, J. Vogel, M. Bonfim, A. Fontaine, Time-resolved X-ray magnetic circular dichroism: a selective probe of magnetization dynamics on nanosecond timescales, in *Spin Dynamics in Confined Magnetic Structures II*, Springer, 2003.
21. A. Yu. Dobin, R. H. Victora, Intrinsic nonlinear ferromagnetic relaxation in thin metallic films, *Phys. Rev. Lett.*, **90**, 167203 (2003).
22. V. Tiberkevich, A. Slavin, Nonlinear phenomenological model of magnetic dissipation for large precession angles: generalization of the Gilbert model, *Phys. Rev. B,* **75**, 014440 (2007).
23. H. T. Nembach, K. L. Livesey, M. P. Kostylev, P. Martin-Pimentel, S. J. Hermsdoerfer, B. Leven, J. Fassbender, B. Hillebrands, Magneto-optical observation of four-wave scattering in a 15-nm $Ni_{81}Fe_{19}$ film during large-angle magnetization precession, *Phys. Rev. B,* **84**, 184413 (2011).
24. C. Kittel, On the theory of ferromagnetic resonance absorption, *Phys. Rev.,* **73**, 155 (1948).
25. J. Smit, H. G. Beljers, Ferromagnetic resonance absorption in $BaFe_{12}O_{19}$, a highly anisotropic crystal, *Philips Res. Rep.,* **10**, 113 (1955).
26. B. Heinrich, Radio frequency techniques, in *Ultrathin Magnetic Structures II*, Springer, 1994.
27. B. Heinrich, Spin relaxation in magnetic metallic layers and multilayers, in *Ultrathin Magnetic Structures III*, Springer, 2005.
28. E. Schloemann, Ferromagnetic resonance in polycrystalline ferrites with large anisotropy, *J. Phys. Chem. Solids,* **6**, 257 (1958).
29. C. Vittoria, *Microwave Properties of Magnetic Films*, World Scientific, 1993.
30. M. Bailleul, Dominik Olligs, C. Fermon, Micromagnetic phase transitions and spin wave excitations in a ferromagnetic stripe, *Phys. Rev. Lett.,* **91**, 137204 (2003).
31. I. Neudecker, G. Woltersdorf, B. Heinrich, T. Okuno, G. Gubbiotti, C. H. Back, Comparison of frequency, field, and time domain ferromagnetic resonance methods, *J. Magn. Magn. Mater.,* **307**, 148 (2006).
32. H. Suhl, Ferromagnetic resonance in nickel ferrite between one and two kilomegacycles, *Phys. Rev.,* **97**, 555 (1955).
33. B. Heinrich, J. F. Cochran, R. Hasegawa, FMR linebroadening in metals due to two-magnon scattering, *J. Appl. Phys.,* **57**, 3690 (1985).
34. Z. Celinski, B. Heinrich, Ferromagnetic resonance linewidth of Fe ultrathin films grown on a bcc Cu substrate, *J. Appl. Phys.,* **70**, 5935 (1991).

35. Y. V. Goryunov, N. N. Garif'yanov, G. G. Khaliullin, I. A. Garifullin, L. R. Tagirov, F. Schreiber, T. Muhge, H. Zabel, Magnetic anisotropies of sputtered Fe films on MgO substrates, *Phys. Rev. B,* **52**, 13450 (1995).
36. K. Lenz, H. Wende, W. Kuch, K. Baberschke, K. Nagy, A. Janossy, Two-magnon scattering and viscous Gilbert damping in ultrathin ferromagnets, *Phys. Rev. B,* **73**, 144424 (2006).
37. J. Linder, I. Barsukov, C. Raeder, C. Hassel, O. Posth, R. Meckenstock, P. Landros, D. L. Mills, Two-magnon damping in thin films in case of canted magnetization: theory versus experiment, *Phys. Rev. B,* **80**, 224421 (2009).
38. S. V. Vonsovskii, *Ferromagnetic Resonance*, Pergamon Press, 1966.
39. T. D. Rossing, Resonance linewidth and anisotropy variation in thin films, *J. Appl. Phys.,* **34**, 995 (1963).
40. R. D. McMichael, D. J. Twisselmann, A. Kunz, Localized ferromagnetic resonance in inhomogeneous thin films, *Phys. Rev. Lett.,* **90**, 227601 (2003).
41. H. Suhl, Theory of the magnetic damping constant, *IEEE Trans. Magn.,* **34**, 1834 (1998).
42. T. Kobayashi, R. C. Barker, J. L. Bleustein, A. Yelon, Ferromagnetoelastic resonance in thin films. I. Formal treatment, *Phys. Rev. B,* **7**, 3273 (1973). T. Kobayashi, R. C. Barker, A. Yelon, Ferromagnetoelestic resonance in thin films. II. Application to nickel, *Phys. Rev. B,* **7**, 3286 (1973).
43. H. Bommel, K. Dransfeld, Excitation of hypersonic waves by ferromagnetic resonance, *Phys. Rev. Lett.,* **3**, 83 (1959).
44. M. Weiler, L. Dreher, C. Heeg, H. Huebl, R. Gross, M. S. Brandt, S. T. B. Goennenwein, Elastically driven ferromagnetic resonance in nickel thin films, *Phys. Rev. Lett.,* **106**, 117601 (2011).
45. W. S. Ament, G. T. Rado, Electromagnetic effects of spin wave resonance in ferromagnetic metals, *Phys. Rev.,* **97**, 1558 (1955).
46. J. M. Lock, Eddy current damping in thin metallic ferromagnetic films, Brit. *J. Appl. Phys.,* **17**, 1645 (1966).
47. N. Smith, Eddy current plus spin dynamical limitations to switching speed in thin-film heads, *IEEE Trans. Magn.,* **27**, 4754 (1991).
48. G. M. Sandler, H. N. Bertram, Micromagnetic simulations with eddy currents of rise time in thin film write heads, *J. Appl. Phys.,* **81**, 4513 (1997).
49. I. D. Mayergoyz, C. Serpico, Y. Shimizu, Coupling between eddy currents and Landau-Lifhitz dynamics, *J. Appl. Phys.,* **87**, 5529 (2000).

50. E. Martinez, L. Lopez-Diaz, L. Torres, Nonphenomenological damping constant due to eddy current losses in uniformly magnetized samples, *J. Appl. Phys.*, **99**, 123912 (2006).
51. G. Hrkac, M. Kirschner, F. Dorfbauer, D. Suess, O. Ertl, J. Fidler, T. Schrefl, Influence of eddy currents on the effective damping parameter, *J. Appl. Phys.*, **99**, 08B902 (2006).
52. C. Scheck, L. Cheng, W. E. Bailey, Low damping in epitaxial sputtered iron films, *Appl. Phys. Lett.*, **88**, 252510 (2006).
53. D. S. Rodbell, Ferromagnetic resonance absorption linewidth of nickel metal. Evidence for Landau-Lifshitz damping, *Phys. Rev. Lett.*, **13**, 471 (1964).
54. L. L. Hirst, R. E. Prange, Exchange effects in ferromagnetic resonance with nonlocal conductivity, *Phys. Rev.*, **139**, A 892 (1965).
55. D. L. Mills, S. M. Rezende, Spin damping in ultrathin magnetic films, in *Spin Dynamics in Confined Magnetic Structures II*, Springer, 2003.
56. N. S. Almeida, D. L. Mills, Eddy currents and spin excitations in conducting ferromagnetic films, *Phys. Rev. B*, **53**, 12232 (1996).
57. B. Heinrich, D. Fraitová, V. Kamberský, The influence of s-d exchange on relaxation of magnons in metals, *Phys. Status Solidi B*, **23**, 501 (1967).
58. V. Kambersky, On the Landau-Lifshitz relaxation in ferromagnetic metals, *Can. J. Phys.*, **48**, 2906 (1970).
59. K. Gilmore, Y. U. Idzerda, M. D. Stiles, Identification of the dominant precession-damping mechanism in Fe, Co, and Ni by first-principles calculations, *Phys. Rev. Lett.*, **99**, 027204 (2007).
60. K. Gilmore, Y. U. Idzerda, M. D. Stiles, Spin-orbit precession damping in transition metal ferromagnets, *J. Appl. Phys.*, **103**, 07D303 (2008).
61. V. Kamberský, Spin-orbital Gilbert damping in common magnetic metals, *Phys. Rev. B*, **76**, 134416 (2007).
62. C. Liu, C. K. A. Mewes, M. Chshiev, T. Mewes, W. H. Butler, Origin of low Gilbert damping in half metals, *Appl. Phys. Lett.*, **95**, 022509 (2009).
63. J. Kuneš, V. Kamberský, First-principles investigation of the damping of fast magnetization precession in ferromagnetic 3d metals, *Phys. Rev. B*, **65**, 212411 (2002).
64. D. Steiauf, M. Fähnle, Damping of spin dynamics in nanostructures: an ab initio study, *Phys. Rev. B*, **72**, 064450 (2005).
65. B. Heinrich, J. D. Meredith, J. F. Cochran, Wave number and temperature dependent Landau-Lifshitz damping in nickel, *J. Appl. Phys.*, **50**, 7726 (1979).

66. J. F. Cochran, B. Heinrich, Microwave transmission through ferromagnetic metals, *IEEE Trans. Magn.*, **16**, 660 (1980).
67. B. Heinrich, G. Woltersdorf, R. Urban, O. Mosendz, G. Schmidt, P. Bach, L. Molenkamp, E. Rozenberg, Magnetic properties of NiMnSb(001) films grown on InGaAs/InP(001), *J. Appl. Phys.*, **95**, 7462 (2004).
68. T. Kubota, S. Tsunegi, M. Oogane, S. Mizukami, T. Miyazaki, H. Naganuma, T. Ando, Half-metallicity and Gilbert damping constant in $Co_2Fe_xMn_{1-x}Si$ Heusler alloys depending on the film composition, *Appl. Phys. Lett.*, **94**, 122504 (2009).
69. M. Oogane, S. Mizukami, Tunnel magnetoresistance effect and magnetic damping in half-metallic Heusler alloys, *Philos. Trans. R. Soc. A*, **369**, 3037 (2011).
70. L. Ritchie, G. Xiao, Y. Ji, T. Y. Chen, C. L. Chien, M. Zhang, J. Chen, Z. Liu, G. Wu, X. X. Zhang, Magnetic, structural, and transport properties of the Heusler alloys Co_2MnSi and NiMnSb, *Phys. Rev. B*, **68**, 104430 (2003).
71. M. Fähnle, D. Steiauf, J. Seib, The Gilbert equation revisited: anisotropic and nonlocal damping of magnetization dynamics, *J. Phys. D: Appl. Phys.*, **41**, 164014 (2008).
72. K. Gilmore, M. D. Stiles, Anisotropic damping of the magnetization dynamics in Ni, Co and Fe, *Phys. Rev. B*, **81**, 174414 (2010).
73. M. Fähnle, D. Steiauf, C. Illg, Generalized Gilbert equation including inertial damping, derivation from an extended breathing Fermi surface model, *Phys. Rev. B*, **84**, 172403 (2011).
74. S. Zhang, S. S.-L. Zhang, Generalization of the Landau-Lifshitz-Gilbert equation for conducting ferromagnets, *Phys. Rev. Lett.*, **102**, 086601 (2009).
75. G. Woltersdorf, B. Heinrich, Two-magnon scattering in a self-assembled nanoscale network of misfit dislocations, *Phys. Rev. B*, **69**, 184417 (2004).
76. Y. Tserkovnyak, A. Brataas, G. E. W. Bauer, Enhanced Gilbert damping in thin ferromagnetic films, *Phys. Rev. Lett.*, **88**, 117601 (2002).
77. Y. Tserkovnyak, A. Brataas, G. E. W. Bauer, Spin pumping and magnetization dynamics in metallic multilayers, *Phys. Rev. B*, **66**, 224403 (2002).
78. Y. Tserkovnyak, A. Brataas, G. E. W. Bauer, B. I. Halperin, Nonlocal magnetization dynamics in ferromagnetic heterostructures, *Rev. Mod. Phys.*, **77**, 1375 (2005).
79. K. Xia, P. J. Kelly, G. E. W. Bauer, A. Brataas, I. Turek, Spin torques in ferromagnetic/normal-metal structures, *Phys. Rev. B*, **66**, 220401(R) (2002).

80. S. Mizukami, Y. Ando, T. Miyazaki, Ferromagnetic resonance linewidth for NM/80NiFe/NM films (NM=Cu, Ta, Pd and Pt), *J. Magn. Magn. Mater.*, **226**, 1640 (2001).

81. T. Gerrits, M. L. Schneider, T. J. Silva, Enhanced ferromagnetic damping in permalloy/Cu bilayers, *J. Appl. Phys.*, **99**, 023901 (2005).

82. B. K. Kuanr, A. V. Kuanr, R. E. Camley, Z. Celinski, Magnetization relaxation in sputtered thin Fe films: an FMR study, *IEEE Trans. Magn.*, **42**, 2930 (2006).

83. H. Lee, L. Wen, M. Pathak, P. Janssen, P. LeClair, C. Alexander, C. K. A. Mewes, T. Mewes, Spin pumping in $Co_{56}Fe_{24}B_{20}$ multilayer systems, *J. Phys. D: Appl. Phys.*, **41**, 215001 (2008).

84. W. M. Saslow, Spin pumping of current in non-uniform conducting magnets, *Phys. Rev. B*, **76**, 184434 (2007).

85. A. Brataas, Y. Tserkovnyak, G. E. W. Bauer, P. J. Kelly, Spin pumping and spin transfer, http://arxiv.org/abs/1108.0385v3.

86. A. Brataas, A. D. Kent, H. Ohno, Current-induced torques in magnetic materials, *Nat. Mater.*, **11**, 372 (2012).

87. A. Brataas, Y. Tserkovnyak, G. E. W. Bauer, B. I. Halperin, Spin battery operated by ferromagnetic resonance, *Phys. Rev. B*, **66**, 060404(R) (2002).

88. A. M. Clogston, H. Suhl, L. R. Walker, P. W. Anderson, Ferromagnetic resonance line width in insulating materials, *J. Phys. Chem. Solids*, **1**, 129 (1956).

89. H. B. Callen, A ferromagnetic dynamical equation, *J. Phys. Chem. Solids*, **4**, 256 (1958).

90. M. Sparks, R. Loudon, C. Kittel, Ferromagnetic relaxation. I. Theory of the relaxation of the uniform precession and the degenerate spectrum in insulators at low temperatures, *Phys. Rev.*, **122**, 791 (1961).

91. M. J. Hurben, C. E. Patton, Theory of two magnon scattering microwave relaxation and ferromagnetic resonance linewidth in magnetic thin films, *J. Appl. Phys.*, **83**, 4344 (1998).

92. R. D. McMichael, M. D. Stiles, P. J. Chen, W. F. Egelhoff, Ferromagnetic resonance linewidth in thin films coupled to NiO, *J. Appl. Phys.*, **83**, 7037 (1998).

93. R. Arias, D. L. Mills, Extrinsic contributions to the ferromagnetic resonance response of ultrathin films, *Phys. Rev. B*, **60**, 7395 (1999).

94. R. Arias, D. L. Mills, Extrinsic contributions to the ferromagnetic resonance response of ultrathin films, *J. Appl. Phys.*, **87**, 5455 (2000).

95. J. B. S. Mendes, L. H. Vilela-Leao, S. M. Rezende, A. Azevedo, Possible interplay between intrinsic and extrinsic ferromagnetic damping mechanisms, *IEEE Trans. Magn.*, **46**, 2293 (2010).
96. D. L. Mills, R. Arias, The damping of spin motions in ultrathin films: is the Landau-Lifshitz-Gilbert phenomenology applicable?, *Phys. B,* **384**, 147 (2006).
97. S. M. Rezende, A. Azevedo, M. A. Lucena, F. M. de Aguiar, Anomalous spin-wave damping in exchange-biased films, *Phys. Rev. B,* **63**, 214418 (2001).
98. J. Lindner, K. Lenz, E. Kosubek, K. Baberschke, D. Spoddig, R. Meckenstock, J. Pelzl, Z. Frait, D. L. Mills, Non-Gilbert-type damping of the magnetic relaxation in ultrathin ferromagnets: importance of magnon-magnon scattering, *Phys. Rev. B,* **68**, 60102(R) (2003).
99. M. Buchmeier, D. E. Bürgler, P. A. Grünberg, C. M. Schneider, R. Meijers, R. Calarco, C. Raeder, M. Farle, Anisotropic FMR-linewidth of triple-domain Fe layers on hexagonal GaN(0001), *Phys. Status Solidi A,* **203**, 1567 (2006).
100. Kh. Zakeri, J. Lindner, I. Barsukov, R. Meckenstock, M. Farle, U. von Hörsten, H. Wende, W. Keune, J. Rocker, S. S. Kalarickal, K. Lenz, W. Kuch, K. Baberschke, Z. Frait, Spin dynamics in ferromagnets: Gilbert damping and two-magnon scattering, *Phys. Rev. B,* **76**, 104416 (2007) [see also corresponding erratum].
101. H. Lee, Y.-H. A. Wang, C. K. A. Mewes, W. H. Butler, T. Mewes, S. Maat, B. York, M. J. Carey, J. R. Childress, Magnetization relaxation and structure of CoFeGe alloys, *Appl. Phys. Lett.,* **95**, 082502 (2009).
102. I. Barsukov, F. M. Römer, R. Meckenstock, K. Lenz, J. Lindner, S. Hemken to Krax, A. Banholzer, M. Körner, J. Grebing, J. Fassbender, M. Farle, Frequency dependence of spin relaxation in periodic systems, *Phys. Rev. B,* **84**, 140410(R) (2011).
103. T. Mewes, R. L. Stamps, H. Lee, E. Edwards, M. Bradford, C. K. A. Mewes, Z. Tadisina, S. Gupta, Unidirectional magnetization relaxation in exchange-biased films, *IEEE Magn. Lett.,* **1**, 3500204 (2010).
104. J. H. van Vleck, R. Orbach, Ferrimagnetic resonance of dilute rare-earth doped iron garnets, *Phys. Rev. Lett.,* **11**, 65 (1963).
105. J. F. Dillon, J. W. Nielsen, Effects of rare earth impurities on ferrimagnetic resonance in yttrium iron garnet, *Phys. Rev. Lett.,* **3**, 30 (1959).
106. J. F. Dillon, L. R. Walker, Ferrimagnetic resonance in rare-earth doped yttrium iron garnet. II. Terbium substitution, *Phys. Rev.,* **124**, 1401 (1961).

107. J. F. Dillon, Ferrimagnetic resonance in rare-earth-doped yttrium iron garnet. III. Linewidth, *Phys. Rev.,* **127**, 1495 (1962).

108. H. T. Nembach, T. J. Silva, J. M. Shaw, M. L. Schneider, M. J. Carey, S. Maat, J. R. Childress, Perpendicular ferromagnetic resonance measurements of damping and Landeí g-factor in sputtered $(Co_2Mn)_{1-x}Ge_x$ thin films, *Phys. Rev. B,* **84**, 054424 (2011).

109. G. Woltersdorf, M. Kiessling, G. Meyer, J.-U. Thiele, C. H. Back, Damping by slow relaxing rare earth impurities in $Ni_{80}Fe_{20}$, *Phys. Rev. Lett.,* **102**, 257602 (2009).

110. R. D. McMichael, C. G. Lee, M. D. Stiles, F. G. Serpa, P. J. Chen, W. F. Egelhoff, Exchange bias relaxation in CoO-biased films, *J. Appl. Phys.,* **87**, 6406 (2000).

111. M. Gloanec, S. Rioual, B. Lescop, R. Zuberek, R. Szymczak, P. Aleshkevych, B. Rouvellou, Temperature dependence of exchange bias in NiFe/FeMn bilayers, *Phys. Rev. B,* **82**, 144433 (2010).

112. P. Lubitz, M. Rubinstein, J. J. Krebs, S.-F. Cheng, Frequency and temperature dependence of ferromagnetic linewidth in exchange biased permalloy, *J. Appl. Phys.,* **89**, 6901 (2001).

113. J. Dubowik, F. Stobiecki, I. Gościańska, Y. P. Lee, A. Paetzold K. Röll, Temperature dependence of ferromagnetic resonance in permalloy/NiO exchange-biased films, *Eur. Phys. J. B,* **45**, 283 (2005).

114. A. M. Clogston, Relaxation phenomena in ferrites, *Bell Syst. Tech. J.,* **34**, 739 (1955).

115. M. Sparks, Effect of impurities on the microwave properties of yttrium iron garnet, *J. Appl. Phys.,* **38**, 1031 (1967).

116. R. Orbach, Spin-lattice relaxation time of rare-earth ions in strong exchange fields, *J. Appl. Phys.,* **33**, 2144 (1962).

117. B. H. Clarke, K. Tweedale, R. W. Teale, Rare-earth ion relaxation time and G tensor in rare-earth-doped yttrium iron garnet. I. Ytterbium, *Phys. Rev.,* **139**, A1933 (1965).

118. M. Gloanec, S. Rioual, B. Lescop, R. Zuberek, R. Szymczak, P. Aleshkevych, B. Rouvellou, Dynamical effect in measurement of the exchange-bias field: a consequence of the slow-relaxer mechanism, *Phys. Rev. B,* **80**, 220404(R) (2009).

Chapter 5

Nonlinear Dynamic Properties of Two-Dimensional Arrays of Magnetic Nanodots

Yuri Kobljanskyj,[a] Denys Slobodianiuk,[a] Gennady Melkov,[a] Konstantin Guslienko,[b,c] Valentyn Novosad,[d] Samuel Bader,[d] Michael Kostylev,[e] and Andrei Slavin[f]

[a]*Faculty of Radiophysics, Taras Shevchenko National University of Kyiv, Kyiv 01601, Ukraine*
[b]*Depto. Física de Materiales, Universidad del País Vasco, UPV/EHU, 20018 San Sebastian, Spain*
[c]*KERBASQUE, Basque Foundation for Science, 48011 Bilbao, Spain*
[d]*Materials Science Division, Argonne National Laboratory, Argonne, IL 60439, USA*
[e]*University of Western Australia, Crawley, WA 6009, Australia*
[f]*Department of Physics, Oakland University, Rochester, MI 48309, USA*
slavin@oakland.edu

When magnetic elements are reduced to a submicrometer size the spin-wave (SW) spectra of these elements are modified substantially. This modification affects not only the linear, but also the nonlinear properties of magnetic elements. In particular, in a flat cylindrical magnetic dot of a radius R the long-wavelength SWs having the wave number $k < 1/R$ are excluded from the SW spectrum and the frequency degeneracy between the quasi-uniform mode of the ferromagnetic resonance (FMR) and higher SW modes

Handbook of Nanomagnetism: Applications and Tools
Edited by Rosa A. Lukaszew

ISBN 978-981-4613-04-0 (Hardcover), 978-981-4613-05-7 (eBook)
www.panstanford.com

could be removed. This spectral modification qualitatively changes all the nonlinear dynamic properties of the dots, as they would not be susceptible to the second-order (four-wave) nonlinear processes that limit the FMR precession amplitude and, consequently, limit the efficiency of all the other nonlinear processes involving a quasi-uniform FMR mode. This effect is demonstrated experimentally in the example of parametric generation of a subharmonic of an external microwave signal in a two-dimensional array of Permalloy nanodots. Due to the reduction of the dot sizes the characteristic lifetime of the electromagnetic radiation at the subharmonic frequency originated from the parametrically excited FMR mode is increased by *2 orders* of magnitude, compared to the case of a continuous magnetic film, where this lifetime is limited by the four-wave relaxation processes involving SWs degenerate in frequency with the quasi-uniform FMR mode.

5.1 Introduction

When the size of a magnetic particle gets smaller—down to a submicron size—all the linear properties of the particle are modified, including the ground state of the particle's magnetization that can be either uniform or vortex like [1].

The spin-wave (SW) spectrum of a magnetic particle is also modified, and if we consider a cylindrical magnetic dot of the radius R and thickness L, we see that with the decrease of size the exchange interaction begins to play a dominant role, and the number of long-wavelength excitations in the SW spectrum of the dot is decreased.

Figure 5.1 demonstrates the qualitative picture of the SW spectrum modification from a continuous magnetic film to a cylindrical magnetic dot of a decreasing radius R. First of all, it is clear from Fig. 5.1 that the frequency of the ferromagnetic resonance (FMR) ω_0 of the magnetic particle, corresponding to spatially uniform precession of magnetization with the wave number $k = 0$, is independent of the absolute sizes of the particle. The FMR frequency can be evaluated using the model of an equivalent ellipsoid for which the effective demagnetization factors N are the functions of the aspect ratio of the particle [2]. For a particle in the form of a thin

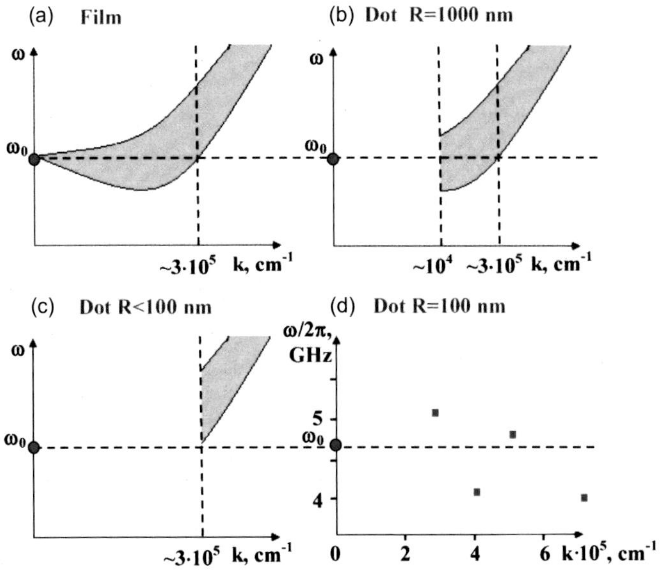

Figure 5.1 Spin-wave spectrum of a magnetic nanodot. The qualitative picture of the spin-wave spectrum (a) of a continuous magnetic film, (b) of a thin magnetic dot having radius $R = 1000$ nm and thickness $L << R$, and (c) of a thin dot having $R < 100$ nm ($L << R$). The angular frequency of the quasi-uniform FMR mode is denoted as ω_0. Frame (d) shows the results (■) of a quantitative calculation (using the formalism [4]) of the frequencies of all the spin-wave modes in an in-plane magnetized (bias magnetic field $H_0 = 300$ Oe) cylindrical Permalloy dot of radius $R = 100$ nm and thickness $L = 10$ nm that differ from the frequency ω_0 of the FMR mode by less than 500 MHz.

disk ($L << R$) situated in the *xz* plane and magnetized to saturation along the *z* axis we get [2]

$$\omega_0 = \gamma\sqrt{H_0[H_0 + (N_y - N_z)M_0]}. \tag{5.1}$$

Here γ is the gyromagnetic ratio for the electron spin, H_0 is the magnitude of the bias magnetic field directed along the z axis, $N_x = N_z << N_y$ are the disk demagnetization factors, and M_0 is the saturation magnetization. It is also clear from Fig. 5.1 that with the decrease of the disk radius due to the influence of the boundaries the long-wave oscillations and waves having the in-plane wave number in the interval $0 < k < 1/R$ are eliminated from the SW spectrum of the disk. Thus, the number of SWs and oscillations degenerate in frequency with the FMR mode is gradually decreasing until this degeneracy is completely eliminated for a magnetic dot of a sufficiently small size (see Fig. 5.1c).

The graphs presented in Fig. 5.1 give a qualitative picture of the SW spectrum modification with the decrease of the in-plane size of a magnetic dot. The real SW spectra of in-plane magnetized magnetic dots can be found using [3, 4] and are *discrete* due to the fact that the dot sizes are limited along the all three Cartesian coordinates. With the decrease of the dot sizes the frequency distances between the individual discrete SW modes increase, and also due to the difference between the shape of a real magnetic disk and an ideal ellipsoid the internal static magnetic field in the disk becomes spatially nonuniform. Thus, the FMR mode of the disk becomes only quasi-uniform: the amplitude of the quasi-uniform FMR precession at the disk edge is smaller than at the disk center [5]. This nonuniformity of the FMR mode profile leads to a slight increase of the FMR frequency with the increase of the dot radius R and to the appearance of the "edge modes" [6] in the dot spectrum.

An example of a discrete SW spectrum of an in-plane magnetized ($H_0 = 300$ Oe) Permalloy (Py) dot of radius $R = 100$ nm and thickness $L = 10$ nm (in the following denoted as 100×10) calculated using the formalism [4] is presented in Fig. 5.1d, where positions of all the SW modes with frequencies that differ from the frequency of the FMR mode by less than 500 MHz are shown. It is clear from Fig. 5.1d that for a magnetic dot of this small size the frequency degeneracy between the FMR mode and the higher SW modes can be completely eliminated, and that there are only five modes having frequencies inside the interval ± 500 MHz from the FMR frequency ω_0. Thus, the discretization of the SW spectra in small magnetic dots by removing the mode degeneracy affects the processes of magnon–magnon relaxation and, therefore, should substantially influence not only linear but also *nonlinear* properties on nanostructured magnetic materials.

It is well known that the efficiency of all the nonlinear processes in magnetically ordered substances increases with the increase of the amplitude of the FMR mode, but in bulk samples and continuous magnetic films this amplitude is limited at a rather low level by the four-wave magnon–magnon interaction processes involving SWs having large magnitudes of the wave number and frequencies that are close to the frequency of the FMR mode [2, 7]. As it has been theoretically demonstrated in Ref. [4] (see also Fig. 5.1), in sufficiently small magnetic particles it should be possible to completely eliminate the frequency degeneracy of the FMR modes with short-wave SW modes and, therefore, to substantially increase the possible amplitude of the FMR mode excited in a magnetic particle. As a result, in a small magnetic particle the efficiency of *all the nonlinear processes*, such as the frequency multiplication, rectification, parametric amplification, or/and generation; parametric wave front reversal; and Brillouin light scattering, should be substantially increased.

To check these ideas we will consider below a nonlinear process of parametric excitation of subharmonic SW modes by a microwave parallel pumping [2, 7, 8] in continuous magnetic films and in magnetic dots of submicron sizes. In particular, we will show that due to the size-dependent SW spectrum modifications in a dot it is possible to increase the lifetime of parametrically excited subharmonic electromagnetic radiation from dots by 2 orders of magnitude, compared to a similar process in a continuous magnetic films [9].

5.2 Parametric Excitation of Spin-Wave Modes in a Magnetic Dot

If a pumping microwave electromagnetic field of the magnitude $h_{\mathrm{p}}(t) = h\cos\left(\omega_{\mathrm{p}}t\right)$ is applied to a magnetic dot parallel to the direction of the in-plane bias magnetic field H_0, and the magnitude of this field exceeds a certain mode-dependent threshold $h_{\mathrm{th}}(k)$ (which is a function of the mode wave number k) the pumping excites an SW or oscillation at a subharmonic frequency $\omega_{\mathrm{p}}/2$ [2, 7, 8]. In the resonance case, when the subharmonic frequency coincides with the frequency of the FMR mode ($\omega_0 = \omega_{\mathrm{p}}/2$), the threshold of excitation of a quasi-uniform ($k \simeq 0$) FMR mode is given by [2]

$$h_{\mathrm{th}}^{(0)} = \frac{2\Gamma_0\omega_{\mathrm{p}}}{\gamma^2 M_0\left|N_x - N_y\right|}, \tag{5.2}$$

while the threshold of excitation of the short ($k >> 1/R$) plane SWs is [2]

$$h_{\mathrm{th}}^{(\mathrm{k})} = \frac{2\left[\Gamma_{\mathrm{k}}^2 + (\omega_{\mathrm{p}}/2 - \omega_{\mathrm{k}})^2\right]^{1/2}\omega_{\mathrm{p}}}{4\pi\gamma M_0 \sin^2\theta_{\mathrm{k}}}, \tag{5.3}$$

where Γ_0, Γ_{k} are the relaxation frequencies of the FMR mode and plane SW, respectively, and ω_{k} and θ_{k} are the eigenfrequency of the plane SW and the angle that the wave vector k of this SW makes with the direction of the bias magnetic field H_0, respectively.

In the SW spectrum of a *continuous* magnetic film there are always many SW modes degenerate in frequency with the quasi-uniform FMR mode ($\omega_{\mathrm{k}} = \omega_0 = \omega_{\mathrm{p}}/2$) and also in a continuous film $\gamma^2 M_0 \left|N_x - N_y\right| = 4\pi\gamma^2 M_0$. Thus, the ratio of the thresholds of excitation of the FMR mode and the SW modes in a continuous film is equal to the ratio of their relaxation frequencies $\Gamma_0/\Gamma_{\mathrm{k}}$. It is also known that in bulk samples and continuous films, due to the influence of two-magnon scattering and radiation damping, the plane SW modes having the wave number $k = k_{\mathrm{min}} \sim 10^4\mathrm{cm}^{-1}$ have the lowest threshold of parametric excitation [2].

When the pumping field exceeds the threshold for the FMR excitation $h > h_{\mathrm{th}}(0)$ the spatially uniform subharmonic oscillations of magnetization start to grow exponentially in magnetic dots. In the n-th dot the variable magnetization corresponding to this parametrically excited subharmonic of pumping is perpendicular to the bias magnetic field $m_{0n} \perp H_0$ and has the form

$$m_{0n}(t) = m_{0n} \cos\left[(\omega_{\mathrm{p}}/2)\, t + \varphi_{\mathrm{on}}\right]. \tag{5.4}$$

The variable magnetization in Eq. 5.4 can be a source of electromagnetic radiation from the dot at a subharmonic frequency $\omega_{\mathrm{p}}/2$. The phase of the parametrically excited subharmonic oscillations in Eq. 5.4 assumes randomly one of the two possible values φ_1 or φ_2, which are determined by the phase φ_{p} of the pumping signal and differ by π [10]:

$$\phi_1 = \frac{\phi_{\mathrm{p}}}{2} + \frac{\pi}{4}(0 \text{ phase}),\ \phi_2 = \frac{\phi_{\mathrm{p}}}{2} - \frac{3\pi}{4}(\pi \text{ phase}),\ \phi_1 - \phi_2 = \pi. \tag{5.5}$$

It is necessary to stress an important property of the parametrically excited oscillations: although the phase of each partial oscillation is undetermined, the sum of phases ψ of the two excited modes (both FMR modes or plane SWs) is fixed at being $\pi/2$ ahead of the phase of pumping [7, 8, 10]. Indeed, in accordance with Eq. 5.5 $\psi = \varphi_1 + \varphi_1 = \varphi_2 + \varphi_2 = \varphi_p + \pi/2$ for the excitation of the uniform FMR mode, and $\psi \equiv \varphi_k + \varphi_{-k} = \varphi_p + \pi/2$ for the excitation of plane SWs. If we consider a parametric excitation of the FMR modes in an array of noninteracting magnetic dots (when the distance between the dot edges is of the order of the dot radius) we have to assume that the parametrically excited oscillations in each dot of the array randomly assume one of the two possible phases (see Eq. 5.5). Thus, the total electromagnetic radiation from the dot array at the subharmonic frequency will be close to zero.

It is possible to obtain a net microwave electromagnetic signal of subharmonic frequency from a parametrically pumped array of magnetic dots if an additional synchronizing signal $h_s = h_{s0} \cos(\omega_s t + \varphi_s)$ $(h_{s0} \perp H_0)$ of the frequency ω_s that is close to the FMR frequency ω_0 of the dot is supplied to the array simultaneously with the pumping. After the application of this synchronizing signal all the subharmonic FMR oscillations in all the individual dots assume the same value of the phase that depends on the ϕ_s of the synchronizing signal. If ϕ_s lies in the interval $\varphi_1 - \pi/2 < \varphi_s < \varphi_1 + \pi/2$ the phase ϕ_1 (0 phase) is established, while if $\varphi_2 - \pi/2 < \varphi_s < \varphi_2 + \pi/2$ the phase ϕ_2 (π phase) is established [10]. In such an externally synchronized state the coherent subharmonic radiation from the parametrically pumped dot array can be experimentally observed.

In the case when the synchronizing signal is applied to each nanodot in a dot array independently we will get a "parametron" [11]—the digital logic device where the roles of "zero" and "unity" are played by the magnetization oscillations having the phase values equal to 0 and π.

5.3 Multimode Magnetization Dynamics above the Parametric Threshold

Above the threshold of parametric excitation the amplitudes of both the uniform FMR mode and the short-wave SW mode begin to grow exponentially. The rate of growth is proportional to the supercriticalities $\zeta_{0,\mathrm{k}} = h/h_{\mathrm{th}}^{(0)}, h/h_{\mathrm{th}}^{(\mathrm{k})}$ [7, 10], where the threshold magnetic fields $h_{\mathrm{th}}^{(0)}, h_{\mathrm{th}}^{(\mathrm{k})}$ are given by Eq. 5.2 and Eq. 5.3. However, the exponential growth of the parametrically excited modes is limited by nonlinear four-wave processes [7]. Due to the four-wave processes the mode having the lowest excitation threshold $h_{\mathrm{th}}^{(\mathrm{k})} = h_{\mathrm{th}}^{\mathrm{min}}$ will, with time, suppress all the other pumped mode and ultimately will be the only one to survive [7]. In contrast, all the other modes, having a larger excitation threshold, will decay exponentially until they are completely suppressed [12].

The processes of four-wave interaction between the SW modes leading to the suppression of certain modes and domination of the other modes can be described in the framework of the so-called S theory [7]. Restricting our consideration to the case of only two modes—a uniform FMR mode N_0 and a short-wave SW mode N_{k}—we can write the following system of equations describing the process of parametric excitation of SW modes by pumping and four-wave nonlinear interaction between the modes:

$$\frac{1}{2}\frac{dN_0}{dt} = N_0\left[-\Gamma_0 + Vh_{\mathrm{p}}\sin\psi\right],$$

$$\frac{1}{2}\frac{dN_{\mathrm{k}}}{dt} = N_{\mathrm{k}}\left[-\Gamma_{\mathrm{k}} + Vh_{\mathrm{p}}\sin\psi\right],$$

$$\frac{1}{2}\frac{d\psi}{dt} = Vh_{\mathrm{p}}\cos\psi + S\left(N_0 + N_{\mathrm{k}}\right). \tag{5.6}$$

Here V, S are the coupling coefficients describing the interaction of modes with an external microwave electromagnetic field and with each other by four-wave processes, respectively [2, 7]. In our simple model we assume that both these coefficients are constant and do not depend on the SW wave vector k. ψ is the phase of a SW pair and N_0, N_{k} are the magnon occupation numbers characterizing the intensity of the uniform FMR mode having $k = 0$ and the SW mode with $k \sim R^{-1}$, respectively. In a limiting case of $N_0, N_{\mathrm{k}} >> 1$ there is a simple relation between the magnon numbers N_0, N_{k} and the square amplitudes of the variable magnetization $m_{0,\mathrm{k}}^2$ corresponding to the excited modes [2]

$$N_{0,\mathrm{k}} = m_{0,\mathrm{k}}^2/\left(2\gamma\hbar M_0\right), \tag{5.7}$$

where $\hbar$ is the reduced Plank constant and M_0 is the static saturation magnetization. Depending on the relative values $\Gamma_0/\Gamma_{\mathrm{k}}$ of the mode's relaxation frequencies different evolution scenarios could take place in a parametrically pumped SW system.

First, we shall consider the case when $\Gamma_k < \Gamma_0$. As it was mentioned above, this situation is typical for bulk magnetic samples or magnetic dots with $R > 1$ μm. As one can see in Fig. 5.1a, in this case the FMR mode is degenerate in frequency with the nonuniform SW mode. In such a case a stationary solution of the system Eq. 5.6 has the following form:

$$N_0^{st} = 0$$
$$N_k^{st} = \frac{1}{S}\sqrt{\left(Vh_p\right)^2 - \Gamma_k^2}. \tag{5.8}$$

It follows from Eq. 5.8 that in the stationary regime above the threshold of parametric excitation the FMR mode is completely suppressed due to the four-wave interaction with SWs, and the only surviving mode is the nonuniform SW mode N_k having the lowest parametric threshold [7].

At the beginning of this process, when $N_{0,k} << N_k^{st}$, the amplitudes of the parametrically excited modes grow exponentially from the starting levels N_0^0, N_k^0 respectively:

$$N_0 = N_0^0 \exp\left[2\left(Vh_p - \Gamma_0\right)t\right],$$
$$N_k = N_k^0 \exp\left[2\left(Vh_p - \Gamma_k\right)t\right]. \tag{5.9}$$

When the synchronizing external microwave signal at the FMR frequency is absent the growth of both the uniform FMR mode and the nonuniform SW modes starts from the thermal level $N_0^0 = N_k^0 = N_T$. In such a situation the nonuniform SW mode having the lowest dissipation ($\Gamma_k < \Gamma_0$) will have the highest growth increment and at $t = \tau_{st} \approx 10\ \Gamma_k^{-1}$ [13] will reach a stationary level, completely suppressing the uniform FMR mode in the process (see Eq. 5.8).

However, if in the same case $\Gamma_k < \Gamma_0$ an external synchronizing signal at the FMR frequency is applied to the system at $t = 0$ the system evolution will be different. The spatially uniform synchronizing signal provides a substantially increased starting amplitude N_0^0 of the uniform FMR mode in every dot and defines the initial phase of the FMR oscillations in all the magnetic dots, thus creating a phase-synchronized coherent state of FMR oscillations of magnetization with the amplitude that substantially exceeds the thermal level $N_0^0 > N_{\mathrm{T}}$. At the same time, the amplitudes of all the other SW modes remain close to the thermal level N_{T}. Obviously, in the absence of the synchronizing signal the initial phases of oscillations in each dot will randomly assume one of the two possible phases given by Eq. 5.5 and no coherent signal at the FMR frequency will be observed.

After the parametric pumping is switched on at $t = 0$ the synchronized FMR oscillations in every magnetic dot will dominate all the other SW oscillations and a coherent electromagnetic radiation at the FMR frequency from the magnetic dot array will be detected. The duration of this radiation strongly depends on the interaction between the FMR mode and the higher SW modes.

In the considered case $\Gamma_{\mathrm{k}} < \Gamma_0$, corresponding to the situation existing in bulk magnetic materials or magnetic dots with $R > 1$ μm, when the initial amplitude N_0^0 of the uniform FMR mode is enhanced by a synchronizing signal this mode will be dominating, but only for a *limited period* of time. As it was explained above, at $t = \tau_{\mathrm{st}}$ the nonuniform SW mode having the lowest dissipation will reach the stationary state determined by Eq. 5.8, and will suppress all the other SW modes, including the uniform FMR mode. Thus, if the duration of the pumping pulse τ_{p} is smaller than τ_{st}, the duration of the electromagnetic radiation from the magnetic dot array caused by the synchronized FMR oscillations in individual dots will be, also, about τ_{p}. In the opposite limiting case $\tau_{\mathrm{p}} > \tau_{\mathrm{st}}$ the duration of the electromagnetic radiation from the dot array will be about τ_{st} due to the suppression of the uniform FMR mode in each dot at $t > \tau_{\mathrm{st}}$ caused by the four-wave interaction with the nonuniform mode having the lowest dissipation. Depending on the dot parameters this duration τ_{st} varies between tens and hundreds of nanoseconds.

Now, let us consider the case of "small" magnetic dotes having $R < 1$ μm. In this case the SW modes with $0 < k < 1/R$ are eliminated from the SW spectrum of a dot (see Fig. 5.1c,d). Thus, the uniform FMR mode will become dominating, having the smallest dissipation parameter: $\Gamma_0 < \Gamma_{\mathrm{k}}$. In such a situation the stationary amplitude of the FMR mode will be nonzero while all the other SW oscillations will be suppressed due to the above described four-wave interaction processes:

$$N_0^{\mathrm{st}} = \frac{1}{S}\sqrt{\left(Vh_{\mathrm{p}}\right)^2 - \Gamma_0^2},$$
$$N_{\mathrm{k}}^{\mathrm{st}} = 0. \tag{5.10}$$

Obviously, in such a case the uniform FMR mode in each dot will be the source of a subharmonic microwave electromagnetic radiation.

However, without a synchronizing external signal at the FMR frequency the total radiation from a dot array will be almost absent as in a previous case due to the random phases of these oscillations in different dots (see Eq. 5.5). The application of a synchronizing signal creates a phase synchronization of FMR oscillations in all the dots in the array, resulting in the appearance of a substantial stationary electromagnetic radiation at the subharmonic of the pumping frequency $\omega_{\mathrm{p}}/2 = \omega_0$. In this case one could expect that the emitted microwave power from the magnetic dot array will remain constant during the whole action of the pumping pulse τ_{p} that can be microseconds long. The power of this subharmonic radiation will be proportional to the pumping power, while the phase of this radiation will be determined by the phase of the synchronizing signal.

We would like stress, that the appearance of the microwave radiation at subharmonic frequency $\omega_{\mathrm{p}}/2$, having a relatively long duration determined by the duration of the pumping pulse, requires two conditions. First, the dots should be sufficiently small, so the SW modes having $k < R^{-1}$ and degenerate in frequency with the FMR mode are eliminated from the SW spectrum of a dot. Second, a synchronizing pulse at the FMR carrier frequency should be applied to the system at $t = 0$ to create phase synchronization between the FMR oscillations parametrically excited in individual magnetic dots by the quasi-uniform microwave pumping.

5.4 Results and Discussion

To prove these ideas we developed the setup shown of Fig. 5.2 for the investigation of parametrically induced subharmonic generation in both patterned and continuous magnetic films. The sample (1) (either a 2D planar array of nanodots or a continuous film) on a dielectric substrate (4) is placed inside an open dielectric resonator

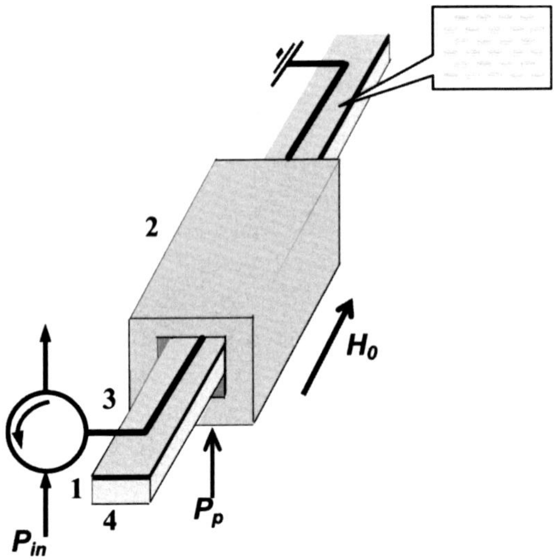

Figure 5.2 Experimental setup. (1) 2D array of Py dots (1.6 ×3.0 mm), (2) open dielectric resonator for the supply of microwave pumping, (3) wire microwave antenna of the diameter of 50 μm, and (4) nonconductive GaAs substrate. P_p is the pumping power, P_{in} is the power of the input synchronizing signal, P_{out} is the power of the output signal created by the subharmonic radiation from the dot array, and H_0 is the in-plane bias magnetic field.

(2) made of a dielectric ceramic with dielectric permittivity $\varepsilon \cong 80$. An external microwave pumping field of the frequency $\omega_p/2\pi$ = 9.4 GHz was applied to the dielectric resonator. The microwave magnetic field h_p of the pumping created in the dielectric resonator was oriented along the plane of the sample and was parallel to the in-plane bias magnetic field H_0. Thus, our experimental geometry was the geometry of a so-called parallel pumping [2, 7]. A short-circuited antenna (3) made of 50 μm wide Cu wire was used to supply a short synchronizing external signal pulse of the power P_{in} to the experimental sample (1) and to receive an output signal P_{out} of electromagnetic radiation from the sample. Both these signals had a carrier frequency of 4.7 GHz corresponding to the subharmonic of the pumping frequency) and were separated using a Y-circulator (see Fig. 5.2). The presence of an external synchronizing signal is vital for the formation of a coherent state in a 2D array of

magnetic nanodots and, therefore, for the detection of the coherent electromagnetic radiation at a subharmonic frequency emitted by the array. It should also be mentioned that the power of the synchronizing signal could be small compared to the power of the microwave parametric pumping.

The experimental samples were either a piece of a continuous Py film or 2D arrays of noninteracting magnetic dots fabricated by patterning of this Py film of the thickness 100 nm deposited on the same nonconductive GaAs substrate of the thickness 0.5 mm. The dots in both investigated arrays had the same dot radius of $R = 1000$ nm and the same interdot distance (distance between the lateral edges of the dots) of $d = 1000$ nm. The first dot array had the dot thickness of $L_1 = 100$ nm (dot array #1), while the second dot array had a substantially smaller dot thickness of $L_2 = 12$ (dot array #2).

At $t = 0$ the samples were subjected to the simultaneous action of a long ($\tau_p = 9$ μs) and powerful (the power $P_p \approx 1-100$ W) pulse of microwave parallel pumping field and a short ($\tau_p = 30$ ns) and relatively weak ($P_{in} \approx 10$ μW) pulse of a synchronizing microwave signal. The output signal P_{out} was received by the antenna (3). As it was expected, the output signal at the antenna (3) appeared only when the pumping power exceeded the threshold power of the excitation of the FMR mode P_p^{th}, which in both investigated dot arrays was around $P_p^{th} \cong 20$ W. The output power P_{out} increased with the increase of the pumping power from the threshold value to the maximum available value of $P_p = 100$ W. The maximum value of P_{out} was obtained by tuning the bias magnetic field H_0 to achieve the resonance condition of the FMR mode ω_0 with the pumping subharmonic $\omega_p/2$: $\omega_0 = \omega_p/2$.

The experimentally measured time dependences of the power of microwave radiation $P_{out}(t)$ with the subharmonic frequency $\omega_0 = \omega_p/2$ for all three samples are presented in Fig. 5.3. As one can see from Fig. 5.3, when the parametric pumping field is switched on at $t = 0$, the output power P_{out} proportional to the intensity of the radiated subharmonic microwave signal starts to increase exponentially. As it was shown previously, the increase of the output power with time is proportional to the supercriticality $\zeta_0 = h/h_{th}^{(0)}$ or in terms of the threshold pumping power

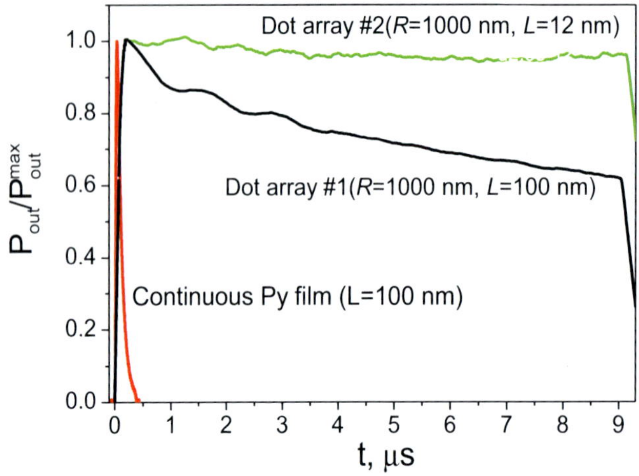

Figure 5.3 Experimental dependence of the normalized output power $P_{\text{out}}/P_{\text{out}}^{\max}$ on the time of action of the microwave pumping. Black line: continuous Py film of thickness $L = 100$ nm, $H_0 = 260$ Oe; red line: dot array #1 ($R = 1000$ nm, $L = 100$ nm), $H_0 = 300$ Oe; green line: dot array #2 ($R = 1000$ nm, $L = 12$ nm), $H_0 = 300$ Oe. Parameters of the pumping and signal pulses: $P_{\text{p}} = 100$ W, $\tau_{\text{p}} = 9$ μs, $\omega_{\text{p}}/2\pi = 9.4$ GHz, $P_{\text{in}} = 10$ μW, $\tau_{\text{in}} = 30$ ns, and $\omega_{\text{p}}/2\pi = 4.7$ GHz.

P_{p}^{th}:$P_{\text{out}}(t) \approx \exp\left[2\Gamma_0\left(\sqrt{P_{\text{p}}/P_{\text{p}}^{\text{th}}} - 1\right)t\right]$, where Γ_0 is relaxation frequency of the uniform FMR mode.

The growth of the output power was limited by the phase mechanism resulting from the four-wave interaction between the parametrically excited SW modes [7], as it was explained above. In the experiment (Fig. 5.3) this level is reached after $t = 200$–300 ns.

Also, we see that the output power in a continuous film grows faster than in both the dot arrays investigated. This is due to the higher threshold of parametric excitation of the SWs in the dot arrays caused by the increase of the relaxation frequency in the dot arrays due to patterning. The temporal evolution of the output power in a continuous film and arrays of nanodots is very different. As explained above, in a continuous film, the influence of the low-threshold SWs with $k \approx 10^4$ cm^{-1} leads to the rapid exponential decrease of the output power P_{out} after the time interval $\tau \approx 200$ ns.

In contrast, in the case of a 2D array of magnetic dots of thickness $L = 100$ nm the decrease of the output power with time is approximately 2 orders of magnitude slower than in the case of a continuous magnetic film (see Fig. 5.3). Also from Fig. 5.3, we see that with the decrease of the dot thickness to $L = 12$ nm, the decrease of the output power with time gets even slower. These are the main experimental results obtained in our paper. Similar results—a slow decrease of the output power with time—were obtained on the dot arrays with the sizes $R \times L$ of 900×50, 1000×40, and 1000×12 (all the sizes are in nanometers).

We attribute the dramatic difference in the temporal evolution of the parametrically excited subharmonic radiation in the patterned dot arrays versus continuous films to the elimination from the dot SW spectrum of SW modes degenerate in frequency with the FMR mode and having $k \leq 10^4$ nm^{-1}. The threshold of parametric excitation of the quasi-uniform FMR mode then becomes lower than that of the other modes. In such a case the theory of parametric excitation [8] predicts that in the stationary regime the amplitude of the parametrically excited FMR mode, and therefore, the output power P_{out} should remain constant during the action of the pumping pulse.

In the experiment shown on Fig. 5.3 we see that for the dot array 1000×12 (sample #2) the output power is, indeed, nearly constant in the whole temporal interval of microwave pumping action, while in the array 1000×100 (sample #1) the output power slowly decreases with time, and this decrease is characterized by a time constant of the order $\tau \approx 10$ μs, which is 2–3 orders of magnitude longer than the typical times of the magnon–magnon relaxation [2, 14]. We believe that slow decrease of the power P_{out} results from the heating of the dots caused by the absorption of the pumped microwave power.

5.5 Heating of Nanodots under the Action of Parametric Pumping

As mentioned above, the decrease of the emitted power P_{out} with time is caused by thermal effects in magnetic dots under the action of microwave pumping. Assuming the dot radius R is much larger than its thickness L, the equation describing dot heating due to absorption of the external microwave pumping field can be derived from the first law of thermodynamics (or conservation of energy), which for a constant volume V of a dot can be written as the equation of heat balance:

$$\Delta U = \Delta Q_{+} - \Delta Q_{-}, \tag{5.11}$$

where the change of internal energy $\Delta U = cV\Delta T$ is equal to the difference between the amount of heat ΔQ_{+} absorbed from the microwave pumping and the amount of heat ΔQ_{-} radiated into the substrate; c is the volume heat capacity of the dot material and ΔT is the change in the dot temperature.

Taking the time derivative of Eq. 5.11 one can obtain the following equation describing temporal evolution of the dot temperature:

$$cV\frac{\partial \Delta T}{\partial t} = P_{\text{abc}} - P_{\text{rad}} \tag{5.12}$$

The power absorbed by the dot P_{abc} is determined by the equation:

$$P_{\text{abc}} = \frac{\partial Q_{+}}{\partial t} = \frac{1}{2}\chi''\omega_{\text{p}}h_{\text{p}}^{2}V, \tag{5.13}$$

where χ'' is the imaginary part of the dimensionless dot magnetic susceptibility, ω_{p} is the angular frequency of the microwave pumping, and h_{p} is the amplitude of the pumping microwave magnetic field that is considered constant along the dot thickness L. This assumption works well if the dot thickness is much smaller than the skin depth at the pumping frequency $\delta\left(\omega_{\text{p}}\right)$.

The power P_{rad} radiated by the dot is proportional to the area of the dot base $A = V/L$ and, in accordance with the Newton's law [15], to the change of the dot temperature ΔT caused by microwave heating:

$$P_{\text{rad}} = \frac{\partial Q_{-}}{\partial t} = \beta\frac{V}{L}\Delta T, \tag{5.14}$$

where β is the coefficient of heat exchange (heat transfer coefficient) between the dot and the substrate measured in $\text{W}/\left(\text{K}\cdot\text{m}^2\right)$. Solving Eqs. 5.12–5.14 the following expression is obtained for the temperature change of the magnetic dot:

$$\Delta T = \frac{1}{2}\omega_{\text{p}}h_{\text{p}}^{2}L\frac{\chi''}{\beta}\left(1 - e^{-t/\tau_{\text{T}}}\right), \tag{5.15}$$

where the characteristic time τ_T of the dot heating is given by:

$$\tau_T = \frac{c}{\beta} L \quad (5.16)$$

It is clear from the solutions of Eq. 5.15 and Eq. 5.16 that the temperature change induced by the microwave heating of the dot and the characteristic time τ_T of this heating are both proportional to the dot thickness L. The typical value of the volume heat capacity for Py $c \approx 4 \cdot 10^6$ J/(K · m^3) and the value of the heat exchange between the substrate and the dot $\beta \approx 4 \cdot 10^4$ W/(K · m^2) [16] lead to the following estimation of the characteristic time of the dot heating: $\tau_T \approx 10$ μs for the dot thickness $L = 100$ nm, which agrees reasonably well with the characteristic time of the power decrease in dots of this thickness shown in Fig. 5.3. Since the change of the dot temperature (Eq. 5.15) is proportional to the dot thickness L the heating-related decrease of the power of microwave subharmonic radiation is much less pronounced for dots of smaller ($L = 12$ nm) thickness (see upper curve in Fig. 5.3).

5.6 Summary

In summary, we studied the influence of sizes of magnetic elements on the nonlinear properties of magnetization oscillations in these elements. In bulk samples and continuous magnetic films the amplitude of the FMR mode is limited by the four-wave interaction processes involving SW modes, having large magnitudes of the wave number, and frequencies that are close to the frequency of the FMR mode. A decreasing of the dot radius has a strong effect on the spatially nonuniform, higher SW modes with wave numbers $k > 0$. Due to influence of boundary conditions at the dot lateral edges the long wavelength part of the SW spectrum is depleted, since the dot can only support SW modes having half-wavelength $\lambda/2 < R$. Thus, the SW modes with wave number $k = 2\pi/\lambda \leq \pi/R$ will be eliminated from the dot SW spectrum (see Fig. 5.1b,c). Also, due to the fact that the dot sizes are finite and small along all the three Cartesian coordinates the SW spectrum of a dot becomes discrete. Thus, the degeneracy of the uniform FMR mode with short-wave SW modes can be completely eliminated in such systems.

The above results allow us to conclude, that for both 2D arrays of magnetic nanodots studied in our experiments the nanostructuring of the magnetic material lead to the exclusion of the of the majority of the four-wave processes of magnon–magnon relaxation, limiting the amplitude of the quasi-uniform FMR mode in a continuous magnetic film and bulk magnetic samples.

This exclusion resulted in a substantial enhancement of the nonlinear properties of the magnetic dot array at the frequency of the main FMR mode. In particular, this nanostructuring resulted in the drastic increase of the characteristic time of parametrically induced subharmonic microwave radiation from the dot array by 2 orders of magnitude in comparison with the case of a continuous magnetic film of a similar thickness. The slow decrease of the radiated subharmonic power observed in Fig. 5.3 is caused by the microwave heating of magnetic dots and can be substantially reduced by reducing the dot thickness (compare curves for the dot arrays of the thickness $L = 100$ nm and $L = 12$ nm shown in Fig. 5.3).

We have proven experimentally that nanostructuring of a magnetic material can substantially enhance the nonlinear dynamic properties of the material. Thus, using nanostructuring, it is possible to develop novel artificial metamaterials with nonlinear microwave properties that are superior to that of magnetic films and traditional bulk magnetic materials. These novel patterned metamaterials can be useful for applications in reciprocal (filters, oscillators), nonreciprocal (isolators, circulators), and nonlinear (detectors, frequency multipliers) microwave signal processing devices operating at high levels of microwave power.

Acknowledgments

This work was supported by the National Science Foundation of the US (grant DMR-1015175), by the DARPA of the US, by the US Army TARDEC, RDECOM, by the State Fund for Fundamental Research of the Ukraine (project #UU34/008), by the MES of Ukraine (grant M/90-2010), by the Australian Research Council, by the US DOE Office of Science (contract DE-AC02-06CH11357), and by the IKERBASQUE (the Basque Foundation for Science).

Konstantin Guslienko also acknowledges partial support by the grants PIB2010US-00153 and FIS2010-20979-C02-01 of the Spanish Ministerio de Economia y Competitividad.

References

1. K. Yu. Guslienko, Magnetic Vortex State Stability, Reversal and Dynamics in Restricted Geometries, *J. Nanoscience and Nanotechnology*, **8**, 2745 (2008).
2. A. G. Gurevich and G. A. Melkov, *Magnetization Oscillations and Waves*, New York: CRC Press, 1996.
3. K. Yu. Guslienko and A. N. Slavin, Spin-waves in cylindrical magnetic dot arrays with in-plane magnetization, *J. Appl. Phys.*, **87**, 6337 (2000).
4. R. Zivieri and R. L. Stamps, Theory of spin wave modes in tangentially magnetized thin cylindrical dots: a variational approach, *Phys. Rev. B*, **73**, 144422 (2006).
5. H. Ulrichs, V. E. Demidov, S. O. Demokritov, and S. Urazhdin, Parametric excitation of eigenmodes in microscopic magnetic dots, *Phys. Rev. B*, **84**, 094401 (2011).
6. J. Jorzick, S. O. Demokritov, B. Hillebrands, M. Bailleul, C. Fermon, K. Yu. Guslienko, A. N. Slavin, D. V. Berkov, and N. L. Gorn, Spin wave wells in nonellipsoidal micrometer size magnetic elements, *Phys. Rev. Lett.*, **88**, 047204 (2002).
7. V. S. L'vov, *Wave Turbulence under Parametric Excitation*, Berlin: Springer-Verlag, 1994.
8. F. R. Morgenthaler, On the possibility of obtaining large amplitude resonance in very thin ferrimagnetic disks, *J. Appl. Phys.*, **33**, 1297 (1962).
9. Y. Kobljanskyj, G. Melkov, K. Guslienko, V. Novosad, S. D. Bader, M. Kostylev, and A. Slavin, Nano-structured magnetic metamaterial with enhanced nonlinear properties, *Sci. Rep.* **2**, 478 (2012).
10. W. H. Louisell, *Coupled Mode and Parametric Electronics*, New York: John Wiley and Sons, 1960.
11. E. Goto and K. F. Loe, *Series in Computer Science, DC Flux Parametron*, Vol. 6, World Scientific, 1986.
12. A. V. Chumak et al., Parametrically stimulated recovery of a microwave signal using standing spin-wave modes of a magnetic film. *Phys. Rev. B.*, **79**, 014405 (2009).

13. S. O. Demokritov, V. E. Demidov, O. Dzyapko, G. A. Melkov, A. A. Serga, B. Hillebrands, and A. N. Slavin, Bose-Einstein condensation of quasi-equlibrium magnons at room temperature under pumping, *Nature*, **443** (430), (2006).
14. V. S. Tiberkevich and A. N. Slavin, Nonlinear phenomenological model of magnetic dissipation for large precession angles: generalization of the Gilbert model, *Phys. Rev. B* **75**, 014440 (2007).
15. H. S. Carslaw and J. C. Jaeger, *Conduction of Heat in Solids, Second Edition*, Oxford: Oxford University Press, 1959.
16. B. Guenin, Transient modeling of a high-power IC package, part 1, *Electron. Cooling*, **17**, 12 (2011).

Chapter 6

Dynamics of Magnetic Nanoparticles and Nanodevices

Stephen E. Russek, Robert J. Usselman, and Eric R. Evarts
National Institute of Standards and Technology, 325 Broadway, Boulder, CO 80305-3337, USA
russek@boulder.nist.gov

Nanoscale magnetic devices and particles are being developed for a wide variety of applications, including magnetic memory, nanoscale sensors, magnetic resonance imaging (MRI) agents, and therapeutic agents. Magnetic nanoparticles are also endogenous to the human body in the form of ferritin. The small volume and anisotropy energies of these structures make their magnetic moments very susceptible to thermal fluctuations. The magnetization dynamics of these structures occur on a wide range of time scales, from thermally activated Arrhenius behavior that may cause an unwanted moment reversal of a magnetic random access memory bit once in 10 years to the 100 GHz precessional resonance of a magnetic nanoparticle

Handbook of Nanomagnetism: Applications and Tools
Edited by Rosa A. Lukaszew

ISBN 978-981-4613-04-0 (Hardcover), 978-981-4613-05-7 (eBook)
www.panstanford.com

in a clinical MRI system. Here we review the stochastic dynamics of magnetic nanoparticles and devices for both single particles and for ensembles. We focus on particles whose dimensions are on the order of 1 nm to 100 nm and have moments on the order of 10 to 100,000 μ_B.

6.1 Introduction

Magnetic nanostructures are being developed for a wide range of applications, including magnetic sensors for hard drives, magnetic random access memory (MRAM) [1, 2], magnetic resonance imaging (MRI) contrast agents [3–5], and therapeutic agents [6]. Magnetic nanoparticles are also endogenous to the human body in the form of ferritin [7, 8]. These magnetic nanostructures may range in size from 1 nm to 100 nm and have moments from 10 to 100,000 μ_B, where μ_B 9.274 $\times$ 10^{-24} Am^2 is the Bohr magneton. Given the small size of these structures, thermal fluctuations are a very important component of the magnetization dynamics [9]. The time scales of the magnetization dynamics vary widely from 100 GHz precessional motion in clinical MRI magnetic fields to the undesirable thermally induced switching of an MRAM bit once in 10 years.

The nanomagnetic structures that we are interested in here are composed mainly of transition metals and transition metal oxides. Three such structures, an MRAM bit, an Fe_3O_4 nanoparticle, and an Fe-8 molecular nanomagnet, are shown in Fig. 6.1. The magnetic moment in these materials is mainly due to the imbalance of spin angular momentum in localized *d* electrons. In metals, the conduction electrons can also be spin-polarized and have spin-dependent properties, for example, the resistance may depend on the electron spin state. The conduction electrons can play a very important role in transporting spin angular momentum in magnetic nanostructures. This conceptual picture, referred to as the *s*-*d* model, dates back to Mott (1936) [10] and has been confirmed for many systems by density functional calculations [11]. The importance of spin transfer by conduction electrons has only recently been recognized [12–14].

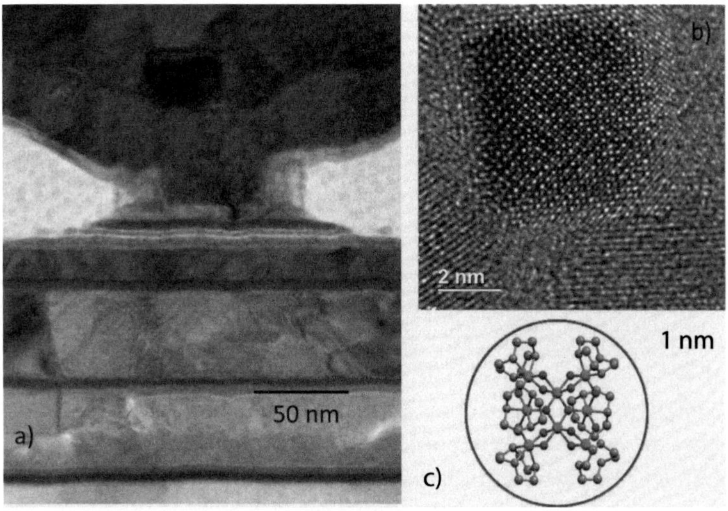

Figure 6.1 (a) TEM micrograph of a magnetic tunnel junction memory bit, (b) TEM micrograph of an iron oxide (Fe_3O_4) nanoparticle (courtesy of Dr. Roy Geiss and Dr. Katherine Rice), and (c) schematic diagram of an Fe-8 $[Fe_8O_2(OH)_{12}(1,4,7\text{-triazacyclononane})_6]Br_8\ 9H_2O$ molecular magnet.

6.2 Stochastic Equations of Motion

The dynamics of a magnetic moment $\vec{\mu}$, whose spins are strongly exchange coupled, can be described by the Landau–Lifshitz–Gilbert (LLG) equation:

$$\frac{d\vec{\mu}}{dt} = -\mu_0|\gamma|\vec{\mu} \times \vec{H}_{\text{eff}} + \frac{\alpha}{|\mu|}\vec{\mu} \times \frac{d\vec{\mu}}{dt} \tag{6.1}$$

where γ is the gyromagnetic ratio and α is a positive dimensionless damping constant. The gyromagnetic ratio is given by $\gamma = g\frac{e}{2m_e}$, where e is the charge on an electron, m_e is the mass of an electron, and the g factor is approximately 2 for the systems that are considered here, giving $\gamma \cong \frac{e}{m_e} = 1.76 \times 10^{11} \frac{\text{rad}}{\text{sT}}$ or $\frac{\gamma}{2\pi} \cong 28$ GHz/T (see Chapter 10 of Ref. [15]). $\vec{H}_{\text{eff}}$ is the effective field, which is the negative gradient of the free energy $U(\vec{\mu}, \vec{H}_{\text{eff}})$ with respect to the magnetic moment

$$\begin{aligned}\mu_0\vec{H}_{\text{eff}} &= -\vec{\nabla}_\mu U = -\vec{\nabla}_\mu(U_{\text{ap}} + U_{\text{ms}} + U_{\text{an}} + U_{\text{ex}} + \ldots) \\ &= \vec{H}_{\text{ap}} + \vec{H}_{\text{ms}} + \vec{H}_{\text{an}} + \vec{H}_{\text{ex}} + \ldots\end{aligned} \tag{6.2}$$

The most important energy terms are the interaction with the applied field (Zeeman energy) $U_{ap} = -\mu_0 \vec{\mu} \cdot \vec{H}_{ap}$, the magnetostatic energy due to dipolar interactions between the spins U_{ms}, the anisotropy energy due to crystalline or interfacial energies U_{an}, and the exchange energy due to spin-dependent quantum-mechanical interactions U_{ex}. The magnetostatic energy of a uniformly magnetized ellipsoidal particle, with principal axes lying along the coordinate directions, is given by $U_{ms} = \frac{1}{2}\mu_0 M_s^2 (N_{xx} m_x^2 + N_{yy} m_y^2 + N_{zz} m_z^2)$ where $\vec{m} = \frac{\vec{\mu}}{\mu}$ is the normalized moment and N_{xx}, N_{yy}, N_{zz} are the shape-dependent demagnetizing factors along the x, y, and z directions, respectively. The sum of the demagnetizing factors is equal to 1. For a sphere the demagnetizing factors are $N_{xx} = N_{yy} = N_{zz} = 1/3$ and there is no preferential direction that minimizes the magnetostatic energy. For an infinitely long cylinder the demagnetizing factors are 1/2 perpendicular to the axis of the cylinder and 0 along the axis so that the moment likes to lie along the axis of the cylinder. For a typical 75 nm $\times$ 50 nm $\times$ 3 nm device layer, shown in Fig. 6.1a, the demagnetizing factors are $N_{xx} = 0.047$, $N_{yy} = 0.072$, and $N_{zz} = 0.881$ and the moment likes to lie in the plane of the film along the long axis of the device. The anisotropy energy is often assumed to have uniaxial ${U_{an}}/{v} = -K_u(\vec{m} \cdot \vec{k}_{an})^2$ or cubic symmetry ${U_{an}}/{v} = K_1[(\vec{m} \cdot \hat{x})^2 (\vec{m} \cdot \hat{y})^2 + (\vec{m} \cdot \hat{z})^2 (\vec{m} \cdot \hat{x})^2 + (\vec{m} \cdot \hat{y})^2 (\vec{m} \cdot \hat{z})^2]$ where K_u and K_1 are the lowest-order uniaxial and cubic anisotropy energy densities and v is the spin cluster volume. If $K_u > 0$ the moment likes to lie along the anisotropy axis $\vec{k}_{an}$, for $K_u < 0$ perpendicular to the anisotropy axis, for $K_1 > 0$ along the <100> axes, and $K_1 < 0$ along the <111> directions. The magnitude of the anisotropy energy is often expressed as an anisotropy field amplitude $\mu_0 H_k = \frac{2K}{M_s}$ where $M_s = \frac{\mu}{v}$ is the saturation magnetization. For a spin cluster with uniaxial anisotropy there is an additional anisotropy field, determined from Eq. 6.2, given by $\vec{H}_{an} = H_k \vec{m} \cdot \vec{k}_{an} \vec{k}_{an}$. Since the effective field is dependent on the moment, the equation of motion is nonlinear. For a recent comprehensive review of dynamics given by the LLG equation see Ref. [16].

The LLG equation can equivalently be written in the form

$$\frac{d\mu}{dt} = -\mu_0 \frac{|\gamma|}{1+\alpha^2} \vec{\mu} \times \vec{H}_{eff} - \frac{\alpha}{|\vec{\mu}|} \frac{|\gamma| \mu_0}{1+\alpha^2} \vec{\mu} \times (\vec{\mu} \times \vec{H}_{eff}) \qquad (6.3)$$

Since this equation dictates that the change in moment is always perpendicular to the existing moment, the magnitude of the moment is conserved, which is appropriate for a single strongly coupled cluster of spins. The trajectories of $\vec{\mathrm{m}} = \frac{\vec{\mu}}{\mu}$ are therefore constrained to the unit sphere. An ensemble of spins that individually obey the LLG equation will have an average moment whose magnitude is not conserved and must be described by a different equation of motion such as the Bloch equations. The first term in Eq. 6.3 describes precession around the effective field, while the second term describes damping and energy loss. Energy flow can be determined by taking the time derivative of the free energy and inserting the equation of motion:

$$\begin{aligned}\frac{dU\left(\vec{\mu}, \vec{\mathrm{H}}_{\mathrm{ap}}\right)}{dt} &= \vec{\nabla}_{\mu} U \cdot \frac{d\vec{\mu}}{dt} + \vec{\nabla}_{\mathrm{Hap}} U \cdot \frac{d\vec{\mathrm{H}}_{\mathrm{ap}}}{dt} \\ &= -\alpha \frac{|\gamma|\, \mu_0}{1+\alpha^2} \left| \vec{\mu} \times \vec{\mathrm{H}}_{\mathrm{eff}} \right|^2 - \mu_0 \vec{\mu} \cdot \frac{d\vec{\mathrm{H}}_{\mathrm{ap}}}{dt} \qquad (6.4)\end{aligned}$$

The precessional term conserves energy since this term is perpendicular to the effective field (energy gradient). The damping term causes the moment to relax in a direction toward the effective field, which moves the system toward a low-energy state. The strength of the relaxation is determined by the dimensionless damping parameter α. The last term in Eq. 6.4 describes the energy transferred into or out of the magnetic system by the applied field.

The damping term describes phenomenologically the interaction of the spin cluster with its environment. The appropriateness of this term depends on the specific nature of the energy loss mechanisms. For instance, in an extended thin film the energy loss to magnons is dependent on the orientation of the moment relative to the film normal [17], which would require an orientation-dependent damping constant. In other systems, the spin cluster may be coupled to a resonant structure that would require incorporation of a frequency-dependent energy loss term. The damping term with a constant α, however, adequately describes the magnetization dynamics in a many systems, including metals where the dominant energy loss is to conduction electrons [18]. The magnitude of the damping constant for materials used in magnetic devices and nanoparticles is typically between 0.005 and 0.05.

The damping term only describes the average energy transfer to the environment. At finite temperatures fluctuations must be included. A simple method to do this is to add a random thermal fluctuation field H_{th} to H_{eff}. H_{th} is typically chosen to be a random Gaussian-distributed field with a root-mean-square (rms) average value of [19]

$$\mu_0 H_{\mathrm{th,rms}} = \sqrt{\frac{2k_{\mathrm{B}}T\alpha}{\mu\gamma\Delta t}} \tag{6.5}$$

where k_{B} is Boltzmann's constant, T is the system temperature, and Δt is the period over which the thermal field is applied. The thermal field/fluctuations increase as the temperature increases, the coupling to the thermal bath increases (α increases), or the particle moment decreases. This stochastic equation describes a classical Wiener process [20] and can be numerically integrated, as shown in Fig. 6.2. Figure 6.2 shows the stochastic dynamics of a 3.5 nm iron oxide particle when placed in a 2 T field along the z direction. The top plot shows the applied field plus the random thermal field given by Eq. 6.5, while the three lower plots show the particle energy, longitudinal, and transverse moments as a function of time. The longitudinal moment, which in this case determines the particle energy since only a Zeeman term is present, wanders about its equilibrium position with a wander rate that depends on α. The transverse moment shows disordered precession about the field.

A probability density function $\rho(\vec{\mu}, t)$ can be defined and a corresponding Fokker–Planck equation can be written.

$$\frac{\partial \rho}{\partial t} = -\vec{\nabla}_{\mu} \cdot \left(\rho \frac{d\vec{\mu}}{dt} - D\vec{\nabla}_{\mu}\rho \right) \tag{6.6}$$

This equation equates the change in local probability density to a divergence of a probability current consisting of a flow given by Eq. 6.3 and a diffusion term arising from thermal interactions. It should be noted that the Fokker–Planck and stochastic LLG approaches give the same results when ensemble averages are compared; however, the Fokker–Planck approach is more amenable to analytic solutions, while the stochastic LLG describes the time evolution of a particular nanostructure and can easily be numerically solved. Both the Fokker–Planck equation and the stochastic

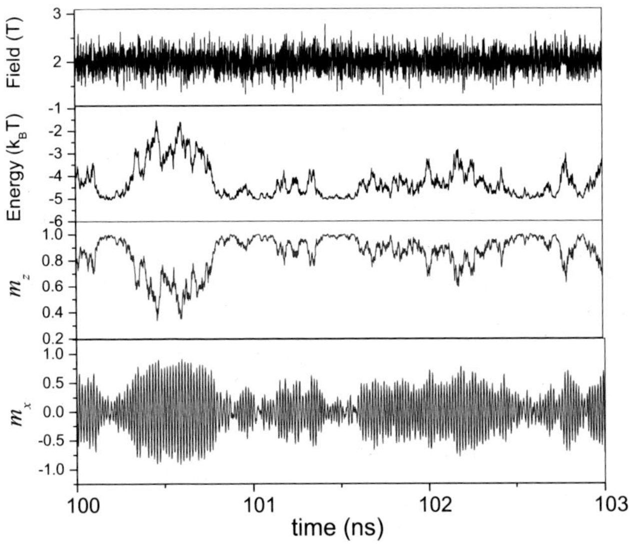

Figure 6.2 Plot of magnetization, energy, and thermal field as a function of time for a 3.5 nm iron oxide nanoparticle at 300 K. $\alpha = 0.01$, $\Delta t \approx 1$ ps.

LLG equation with the thermal field break down when thermal kicks (the sharp jumps in energy seen in Fig. 6.2) are large and on the order of the total magnetic energy. Equation 6.6 assumes that the thermal processes are small and can be modeled with a diffusion term. When the magnetic system gets very small one has to use a different stochastic equation, for example, a jump model, in which a thermal kick can cause the moment to jump anywhere on the phase space sphere [21]. This is the case for electron paramagnetic resonance, for example, a Cu^{+2} spin or a molecular nanomagnet at low fields and high temperatures, where the thermal energy is much larger than the magnetic energy.

Examples of the stochastic dynamics of a 5 nm iron oxide particle with uniaxial 40 kJ/m^3 ($\mu_0 H_k = 0.267$ T) anisotropy are shown in Fig. 6.3 for $T = 0$ K, 20 K, and 300 K. At zero temperature the moment precesses and relaxes back to a low-energy configuration. For systems with low damping, the motion can be thought of as a slow progression along precessional constant-energy surfaces. At moderate temperatures ($T = 20$ K in Fig. 6.3) the magnetization

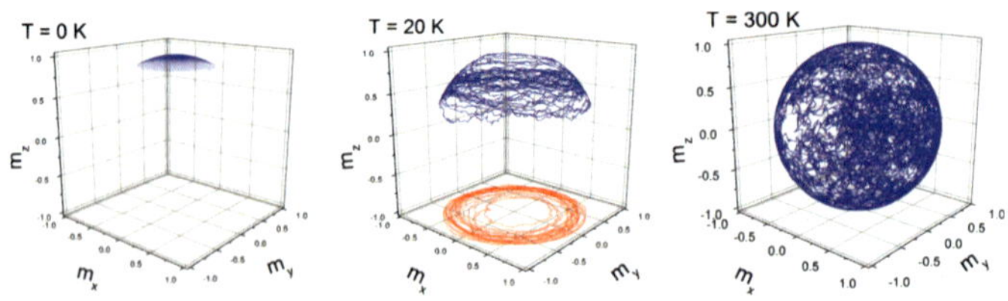

Figure 6.3 Trajectories for a 5 nm iron oxide particle with uniaxial anisotropy along the z axis. At low temperature the moment precesses and relaxes back to a low-energy state, at intermediate temperatures the moment "diffuses" between different precessional orbits close to the energy minimum, and at high temperature the trajectory wanders over all of phase space with slightly more time spent in the low-energy regions along the z axis. The $T = 20$ K plot shows the projection of the orbits in the $m_x - m_y$ plane (red line), which more clearly shows disordered precessional orbits.

wanders about disordered precessional orbits close to an energy minimum. At high temperatures ($T = 300$ K in Fig. 6.3) the moment wanders over the whole unit sphere. For systems with large damping, strong coupling to the environment, or very small size, the thermal kicks (Eq. 6.5) are very large and the precessional motion is barely discernible. In the frequency domain we say that the ferromagnetic resonance (FMR) linewidth becomes very broad.

For a spherical particle with no anisotropy in a magnetic field $\mathbf{B} = \mu_0 \mathbf{H}_{\mathrm{ap}}$ along the z direction, the time average longitudinal magnetization is given by the Langevin function

$$\langle \mu_z(t) \rangle = \mu \left[\coth\left(\frac{\mu B}{k_{\mathrm{B}} T} \right) - \frac{k_{\mathrm{B}} T}{\mu B} \right] \tag{6.7}$$

The time average is equal to the average of the z component of the moment times the Boltzmann factor and is independent of the damping constant. Numerical simulations, shown in Fig. 6.4a, confirm that the stochastic equations of motion do indeed give the correct ensemble average.

One can relate the power spectrum to the imaginary part of the magnetic susceptibility through the fluctuation-dissipation theorem [22, 23]

$$\left\langle m_i(f) m_j(f)^* \right\rangle = \frac{k_{\mathrm{B}} T}{2\pi f \mu_0 M_s^2 V} \chi''_{ij}(f) \tag{6.8}$$

where $m_i(f)$ is the Fourier transform of the ith component of $\vec{m}(t)$. The dimensionless volume susceptibility is defined by

$$m_i = (\chi'_{ij} + i\chi''_{ij})\Delta H_j(f) \tag{6.9}$$

where ΔH is a small perturbative field, for example, microwave excitation in the case of FMR experiments. The averages in Eqs. 6.6 and 6.7 are ensemble averages, which can be either different stochastical motions for identical particles or for an ensemble of particles with a distribution of properties. For identical 3.5 nm particles, the numerically calculated transverse power spectra are shown in Fig. 6.4 for two different values of damping. One observes an FMR peak whose resonant frequency f_{r} increases with applied field and whose width increases with increasing damping. When the thermal energy is small compared to the magnetic energy, at low temperatures or high fields, the complex transverse susceptibility reduces to the deterministic susceptibility of the LLG equation for small perturbations about the equilibrium, which for a particle with a small uniaxial anisotropy field B_{k} in a magnetic field of B_z is given by [15]

$$\chi_{xx} = \frac{\gamma'\mu_0 M_{\mathrm{s}}(\gamma' B_z + i\alpha f)}{f_{\mathrm{r}}^2 - f^2 + 2\alpha f\gamma' i B_z}$$

$$f_{\mathrm{r}} = \gamma'\left(B_z + \frac{1}{2}B_{\mathrm{k}}(3\cos^2(\theta) - 1)\right) \tag{6.10}$$

where γ' is $\gamma/2\pi$ and θ is the angle between the applied field and the particle easy axis. This equation is used to calculate the power spectrum of the 3.5 nm particles in Fig. 6.4b and is plotted as the red solid line. The agreement with the numerical data is quite good at high fields but substantially deviates at low fields when the particles are not saturated. In this case, thermal fluctuations give rise to additional broadening (discussed in the next section) not accounted for by Eq. 6.10. Equation 6.10 predicts, in the simple case of particles whose moment undergoes small excursions around their low-energy states, the intrinsic FMR line width (full-width at half-maximum) should increase linearly with frequency

$$\Delta f = 2\alpha f_{\mathrm{r}}; \quad \Delta B = \frac{2\alpha f_{\mathrm{r}}}{\gamma'} \tag{6.11}$$

For real nanoparticles with complex energy surfaces the magnetic resonance line shape and width will be more complex then

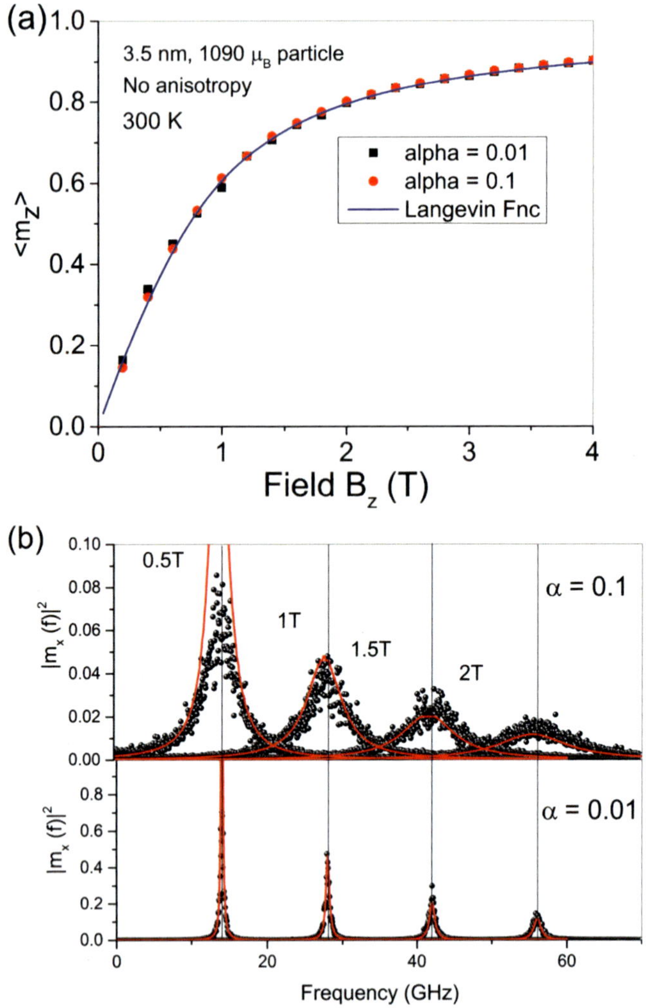

Figure 6.4 (a) Average moment along the magnetic field direction determined from numerical solutions of the stochastic equations of motion of a 3.5 nm iron oxide particle with no anisotropy and predicted equilibrium moment (solid line) and (b) transverse power spectra obtained from the same set of dynamical motions for several magnetic field values and for two different damping constants.

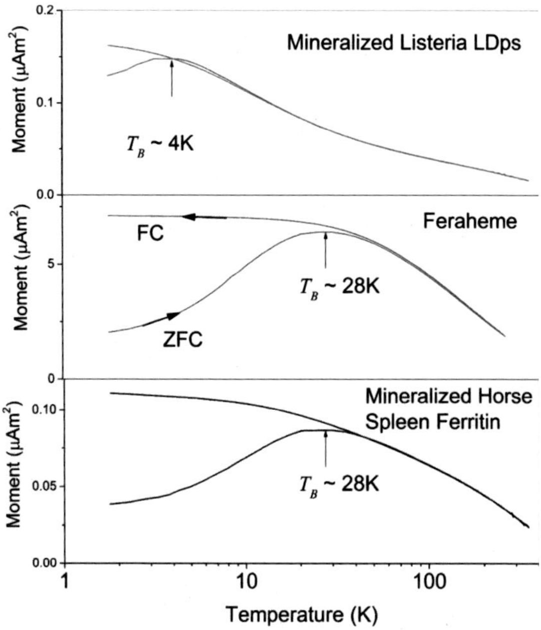

Figure 6.5 Zero-field-cooled (lower curves) and field-cooled (upper curves) moments for three types of iron oxide particles: mineralized *Listeria* Dps ~4 nm, feraheme ~5 nm, and mineralized horse spleen ferritin ~6 nm.

given by Eqs. 6.10 and 6.11. Orbital frequencies will vary as the moments sample different trajectories. Since the equations of motion are nonlinear, the orbits will not have a single frequency component but will contain higher harmonics.

6.3 Dynamics of Superparamagnetic Particles

6.3.1 *Superparamagnetism*

Superparamagnetism refers to fluctuations in spin clusters that can have complex energy surfaces. Unlike a simple paramagnetic species, the magnetic response will be more complex than the Langevin behavior shown in Fig. 6.4a, and the moment can display hysteretic behavior. Examples of hysteresis for several different nanoparticle systems are shown in Fig. 6.5. Here the magnetic moments of nanoparticle ensembles are shown (1) when the system has been cooled in a zero field (zero-field-cooled, ZFC) and then warmed in a small magnetic field and (2) when the particles moments are measured as the sample is cooled in a small magnetic field (field-cooled, FC). At high temperatures, the moments fluctuate and sample all of phase space in the measurement time t_m, and hence the ZFC and FC moments are the same. At low temperatures, the moments stop fluctuating and are trapped in local energy minima that depend on sample history. In the case of ZFC measurements, the moment at low temperatures should be frozen into a disordered low-moment state. The onset temperature of fluctuations is denoted as the blocking temperature T_B. The blocking temperature is a function of the observation time or measurement frequency. Depending on the application, moment fluctuations may be either desirable or undesirable. For biological contrast agents or iron supplement applications, superparamagnetism prevents the iron particles from agglomerating and allows them to retain a stable colloidal solution, whereas for a memory bit fluctuations cause unwanted reversal of the bit state and write errors.

Figure 6.6 shows numerical simulations of a 6 nm × 5 nm × 3 nm superparamagnetic particle below T_B at 5 K, just above the blocking temperature at 20 K, and at 80 K in the paramagnetic region. Here, the magnetostatic shape anisotropy gives an easy axis along the x direction with an energy difference of $(U_x - U_y)/k_B = 57$ K. At low temperature one observes small fluctuations about the local energy minima and a small FMR peak. Just above the blocking temperature one observes switching between energy minima and a large low-frequency peak in the longitudinal susceptibility and a broad FMR peak. At high temperature the system behaves like a paramagnetic particle.

6.3.2 *Fluctuation Broadening*

A small magnetic particle at elevated temperature will sample phase space on a time scale on the order of $t = \tau_0 e^{-\frac{\delta U}{k_B T}}$, where τ_0 is an attempt time on the order of 10^{-12} to 10^{-9} s and δU is the maximum energy difference found over the spherical phase space. For systems with small damping, the dynamics can be viewed as the

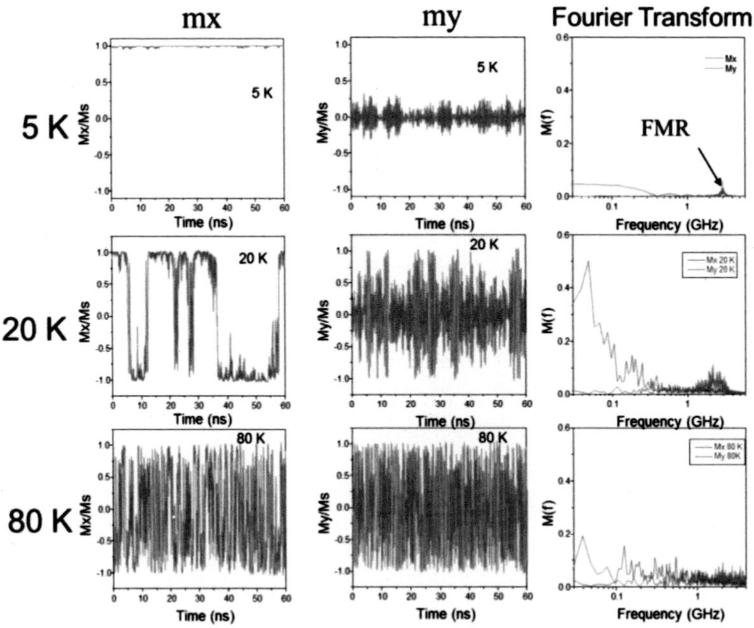

Figure 6.6 Dynamical simulations of longitudinal ($\mathbf{m}_x$) and transverse ($\mathbf{m}_y$) moments of a 6 nm × 5 nm × 3 nm iron oxide particle with only shape anisotropy, $\alpha = 0.01$ and at temperatures below, just above, and far above the blocking temperature. Shown also are the power spectra showing both FMR peak and low-frequency noise.

spin system occupying various precessional orbits in phase space. The susceptibility will have a resonant behavior with a resonance linewidth determined by phase and frequency fluctuations. The magnetic moment thermally hops from one orbit to another or along an orbit. If all of the orbits have the same frequency, for example, when only the Zeeman interaction is present, then the broadening will be due to random changes in phase of the orbit as the system hops along or between orbits. This leads to the intrinsic line width due to damping, as shown in Fig. 6.4b. If the system has other energy terms, such as anisotropy or demagnetization energy, then the frequency of the orbits vary across phase space. A second type of broadening due to thermal fluctuations is then the sampling of many different precessional orbits with different frequencies. This effect is

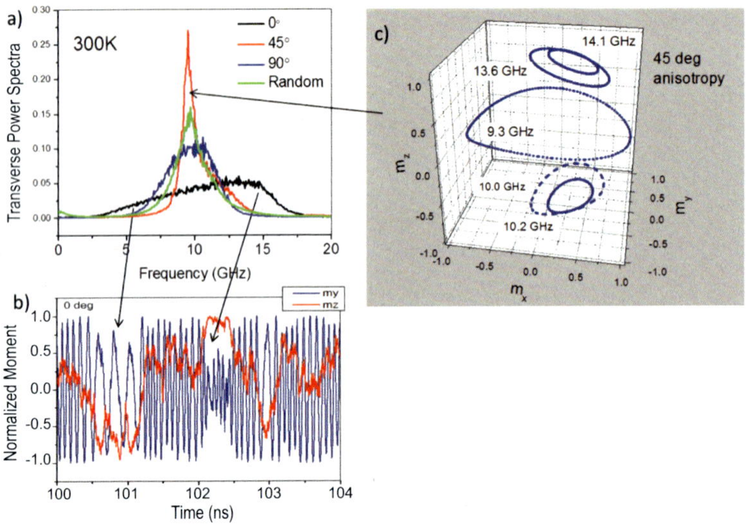

Figure 6.7 (a) Numerical simulations of magnetization power spectra for 3.5 nm iron oxide nanoparticles with uniaxial anisotropy of magnitude $K_u =$ 40 kJ/m^3 for different directions of the applied field relative to the easy axis. The applied field has a magnitude of 0.34 T applied along the z axis. (b) Magnetization dynamics for a particle with the easy axis aligned with the field. The moment precesses at high frequencies when the moment is aligned with the field ($m_z = 1$) and at low frequency when the moment is antialigned ($m_z = -1$). (c) Sample orbits for a particle with the easy axis located 45° from the applied field direction. The predominant equatorial orbits have "g = 2" frequencies.

shown in numerical solutions of the stochastic LLG equation plotted in Fig. 6.7. The transverse power spectra of a 3.5 nm iron oxide particle is shown for different directions of the uniaxial anisotropy axis relative to the applied field of magnitude 0.34 T applied along the z axis. As seen, the power spectra are very broad and asymmetric. For the case where the anisotropy axis lies along the field axis (0° in Fig. 6.7a), the time domain data in Fig. 6.7b clearly shows that the high-frequency component, $f_0 = g\frac{e}{2m_e}\mu_0\left(H_{\text{ap}} + H_{\text{k}}\right)$, occurs when the particle's moment lies along the field/easy axis and the low-frequency components, $f_0 = g\frac{e}{2m_e}\mu_0\left(H_{\text{ap}} - H_{\text{k}}\right)$, arise when the moment lies antiparallel to the field direction. When the moment is perpendicular to the easy axis, the anisotropy does not contribute, $\vec{H}_{an} = H_k \vec{m} \cdot \vec{k}_{an} \vec{k}_{an} = 0$, and the resonant frequency is approximately $f_0 = g \frac{e}{2m_e} \mu_0 H_{ap}$. Here, the apparent g_{eff} value is given by the intrinsic value close to $g = 2$ and these orbits are often referred to as "$g = 2$ orbits." Note that when there is a random anisotropy direction, the green curve in Fig. 6.7a, there is a predominant peak near $g = 2$ due to the fact that there is more phase space perpendicular to an easy axis as opposed to along the easy axis (i.e., the polar region is much smaller than the equatorial region). This effect is shown explicitly in Fig. 6.7c, which shows the orbits for a particle whose anisotropy axis is 45° to the applied field. The equatorial orbits have frequencies near $g_{eff} = 2$.

The anisotropy energy assumed in Fig. 6.7 (+40 kJ/m^3, $\mu_o H_k =$ 0.27 T) is typical for that measured in iron oxide nanoparticles [24, 25]. This energy is much greater than the magnetocrystalline anisotropy of magnetite (–11 kJ/m^3). Given the lack of detailed knowledge about the source of the anisotropy, it is often assumed that the anisotropy is uniaxial. Angle-dependent FMR measurements on field-aligned particles usually show that the resonance frequency is periodic with a period of 180° indicating a predominant uniaxial anisotropy [24], although difficulty aligning the particle easy axes makes this technique not conclusive. The sign of the uniaxial anisotropy is also hard to determine. Here we have assumed that it is positive corresponding to a prolate ellipsoid. If the uniaxial anisotropy energy were negative, corresponding to an oblate ellipsoid, then the spectra would be quite different with a peak at higher frequencies and a low-frequency shoulder, as seen in Fig. 6.8, which shows calculated spectra for 5 nm iron oxide particles with positive and negative uniaxial anisotropy in an applied field of $\mathbf{B}_a = 0.34$ T. While the power spectra, which are proportional to the magnetic susceptibility, are calculated as a function of frequency, the top axis shows the nominal field $\mathbf{B}_n = (\mathbf{B}_a - f/\gamma') + \mathbf{B}_a$, assuming that the susceptibility is a function of $\mathbf{B}_a - f/\gamma'$, as indicated in Eq. 6.10. Plotting versus the nominal field allows a better comparison with experimental data, shown in the next section, which often measure the susceptibility as a function of magnetic field rather than frequency. The asymmetry in the observed FMR spectra is an important source of information on the type of anisotropy present.

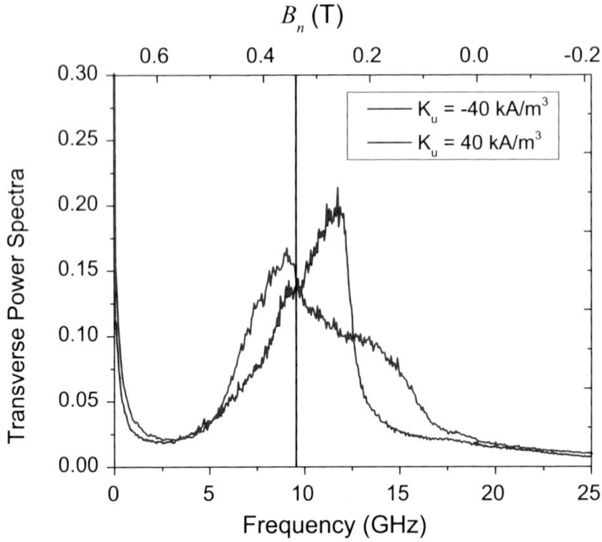

Figure 6.8 Numerical calculations of transverse power spectra for 5 nm iron oxide particles at 20 K in an applied field of $B_a = 0.34$ T for positive and negative uniaxial anisotropy energies.

6.3.3 *Ensemble Effects*

It is difficult to measure a single nanoparticle and measurements are often done on ensembles that have a distribution of properties, including particle size, anisotropy energy density, anisotropy axis, etc. Given the large number of possible ensemble distributions, it is often impossible to uniquely determine particle properties from measured data. Examples of X-band (~10 GHz) magnetic resonance measurements, taken at 300 K, on ensembles of randomly oriented magnetic nanoparticles are shown in Fig. 6.9 (the derivative signals are shown in the plot and the integrated signals in the inset). The data plotted include mineralized *Listeria* Dps protein cages (DNA-binding proteins from starved cells), horse spleen ferritin mineralized with maghemite/magnetite and feraheme, a commercial iron supplement drug. The data show very broad asymmetric resonances similar to the theoretical curves shown in Fig. 6.8. A shoulder on the low field side suggests that the particles have a positive uniaxial anisotropy. There are also additional sharp

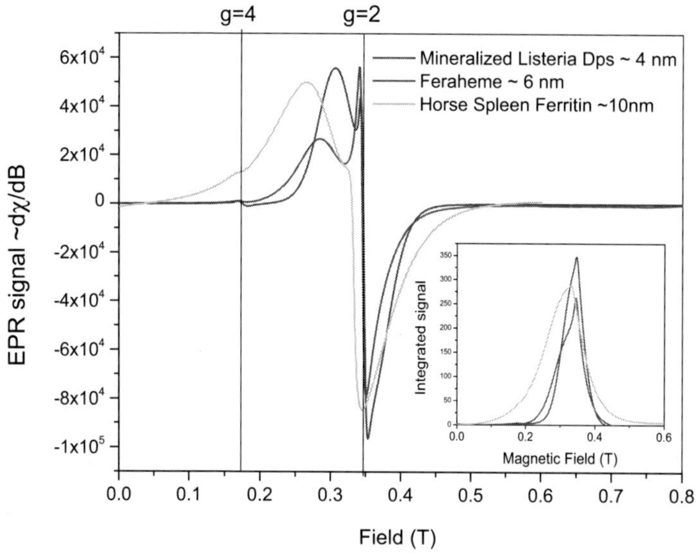

Figure 6.9 Measured X-band (9.7 GHz) magnetic resonance spectra for several different randomly oriented iron oxide nanoparticle samples at $T = 300$ K.

features near $g = 2$ that have been ascribed to surface spins, small particles, or low-anisotropy particles [26]. The $g = 2$ feature becomes more pronounced as the particle size decreases.

Raihker and Stepanov (RS) 1992 have calculated, using a Fokker–Plank approach, the expected susceptibility for an ensemble of particles with either uniaxial or cubic anisotropy with randomly distributed axes. The RS calculation, which results in analytical results, assumes the applied field is much greater than the anisotropy field, which is often not the case. For example, in Fig. 6.9 the anisotropy fields are on the order of 0.25 T compared to the applied field value of 0.34 T. More comprehensive calculations have been obtained by Biasi et al. [27].

The calculated susceptibilities using the RS analytical results, for the case of particles with positive uniaxial anisotropy, are shown in Fig. 6.10 along with magnetic resonance data from 5–6 nm iron oxide particles. At low temperatures, the susceptibility displays a broad "powder pattern" with a width on the order of $2H_k$. Here,

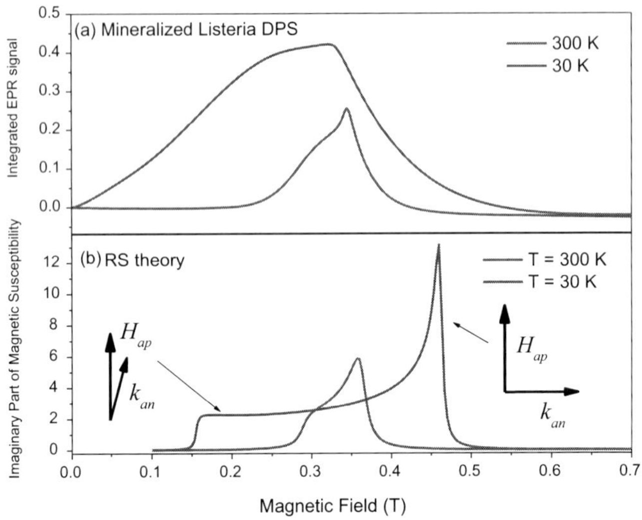

Figure 6.10 (a) Calculated X-band susceptibilities of 5 nm iron oxide particles with 40 kJ/m^3 uniaxial anisotropy energy with random anisotropy axes using the theory of Raikher and Stepanov. As the temperature increases the peak width narrows and centers on $g_{\mathrm{eff}} = 2$. (b) X-band magnetic resonance spectra of mineralized *Listeria* cages with ~5 nm iron oxide cores. The sample was a randomly oriented powder sample.

the moments precess about their low-energy state approximately aligned along the field axis with a resonant frequency dependent on the angle between the applied field and the anisotropy axis. In the limit of large applied fields and low temperatures, the resonance frequency is given by $f_0 = g\frac{e}{4m_e}\mu_0 H_{\mathrm{ap}}\left(3\cos^2(\theta) - 1\right)$, where θ is the angle between the applied field and the easy-axis direction. The susceptibility at high fields in Fig. 6.10b is due to low-frequency orbits, which have $\theta \approx 90^\circ$, while the low-field susceptibility is due to high-frequency orbits, in which $\theta \approx 0^\circ$. Since there are more particles with easy axes perpendicular to the field there is more spectral weight at high fields for a system with positive randomly distributed uniaxial anisotropy.

The observed inhomogeneous line width is much greater than the intrinsic homogenous line width, which for the X-band data shown in Fig. 6.10 is only 3 mT. As the temperature increases, the line width narrows and centers about $g = 2$. At high temperatures the moments fluctuate and sample the entire spherical phase space and the predominant equatorial orbits are at $g = 2$. This narrowing of the inhomogeneous line width is often referred to as "anisotropy melting." The spectra calculated by RS theory, in the high field limit, agree with the data obtained by numerically integrating the stochastic LLG equation. Numerically integrating the stochastic LLG equation, however, provides data over the full field and temperature range and also predicts higher harmonic components that must be present when there is significant magnetic anisotropy.

While the RS and Biasi models can fit the measured data at high temperatures, there is significant divergence at low temperature, indicating that the real ensemble cannot be represented by a set of particles with single uniaxial anisotropy energy with random anisotropy axes.

6.4 Dynamics of Nanoscale Magnetoelectronic Devices

Nanoscale magnetic devices, spin valves or magnetic tunnel junctions, are being used in commercial hard drives as magnetoresistive read sensors and for MRAM. Most of these devices use complex multilayer stacks. Example magnetic tunnel junction and spin-valve stacks are shown in Fig. 6.11. The resistance of these devices is a function of the angle between magnetizations of the two magnetic layers sandwiching the spacer layer. The resistance in the antiparallel state is higher than the parallel state since in the antiparallel state, electrons traversing the device must change from being majority to minority spins with a concomitant change in the potential energy. The change in the potential leads to increased scattering at the interfaces. Magnetoresistance values can range from 5% for a spin valve up to 400% for a high-quality magnetic tunnel junction. Magnetic devices are ideal for studying nanoscale magnetodynamics since the conduction electrons, and hence the device resistance, are strongly coupled to the local magnetic order. One can directly measure the magnetization dynamics by high-bandwidth measurement of the device impedance [28]. At present, these device systems are the only ones in which gigahertz

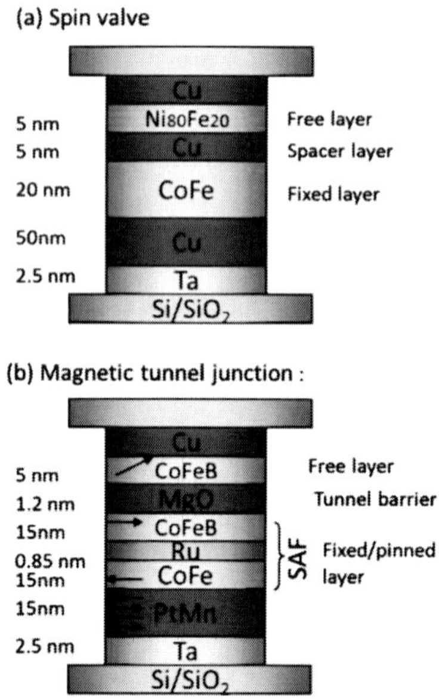

Figure 6.11 (a) Spin-valve stack used for current-perpendicular-to-the-plane devices and (b) magnetic tunnel junction stack.

magnetization dynamics can be measured in real time in sub-100 nm structures.

6.4.1 *Dynamics in Magnetoresistive Sensors*

Most current devices have only one layer, the free layer, which undergoes large dynamical motion. For sensors, the free layer moves in response to the bit field. Commercial hard drive read sensor (year circa 2004) responses are shown in Fig. 6.12. For small devices, the free layer is never really in a quiescent state, and thermal fluctuations are always present, as shown in the magnetic noise spectra in the bottom plot in Fig. 6.12a. This particular device shows a single-domain response, which displays an FMR peak near 4 GHz that moves with an applied transverse magnetic field. For this application it is important to keep the resonant fluctuations above the operation frequency ($\sim$1 GHz).

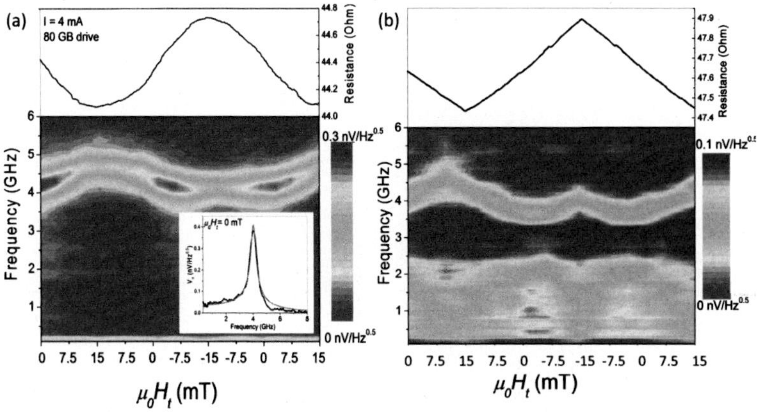

Figure 6.12 (a) Magnetic noise spectra and magnetoresistance of a commercial hard drive read sensor, year circa 2004, showing near-ideal single-domain behavior. (b) Magnetic noise spectra and magnetoresistance of a commercial hard drive read sensor, circa 2004, showing considerable low-frequency noise, in addition to intrinsic FMR spectra.

Many devices show complex thermal fluctuation spectra such as those shown in Fig. 6.12b. This device shows additional thermal fluctuations at frequencies below the FMR frequency, presumably due to defects from poor device fabrication such as a nonuniform easy-axis bias field.

6.4.2 *Magnetic Switching/Memory*

In conventional magnetic field switching of a nanoscale magnetic device, such as an MRAM bit, an applied field causes a reduction of the energy barrier until the moment is thermally activated over the barrier or the barrier completely disappears. The device then relaxes into a new state. Concurrently, the frequency of the main excitation mode goes to zero. Conventional switching therefore occurs when an excitation mode frequency is driven to zero. This is shown explicitly in Fig. 6.13. The top trace shows the device resistance versus the easy-axis magnetic field, which displays spikes when the free layer switches. The bottom spectral plot shows the measured device voltage noise showing the FMR peak going to zero frequency at switching fields of ~7 mT. A different mode of switching

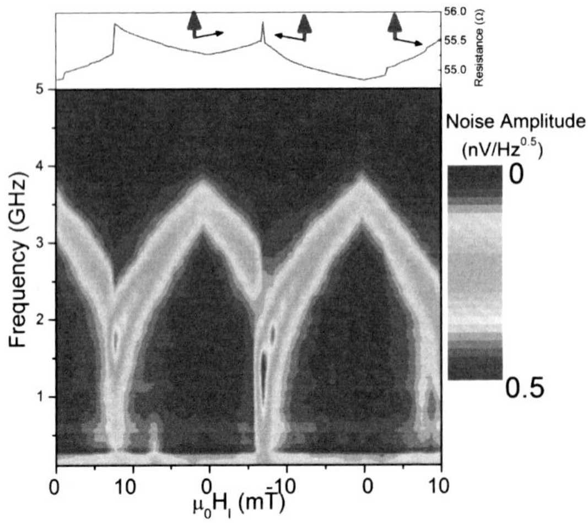

Figure 6.13 Magnetic noise spectra of a spin-valve device as a magnetic field is applied to cause the device to switch states [29].

can be realized by spin transfer effects discussed in the next section. Instead of having a mode whose frequency goes to zero, one has a mode whose line width goes to zero, indicating that the effective damping has gone from positive to negative.

The magnetic noise spectra for a 50 nm × 150 nm magnetic tunnel junction MRAM bit is shown in Fig. 6.14 as a function of a perpendicularly applied magnetic field. Several modes are seen, and the dynamics are more complex than the devices that exhibit single-domain behavior. The MRAM device operates by having the free layer switch between its two in-plane easy-axis low-energy configurations, which have different resistances. To allow for easier and more stable switching additional perpendicular anisotropy is added to partially compensate for the demagnetizing energy, which causes the moment to prefer to line in the plane of the film. The additional perpendicular anisotropy allows the free-layer moment to precess out of the plane of the film and more easily switch. The frequency of the main mode goes to zero at the field (300 mT) where the perpendicular applied field compensates for the net anisotropy field given by $\mu_0(\mathbf{M}_s - \mathbf{H}_{kperp})$. The higher-frequency modes often

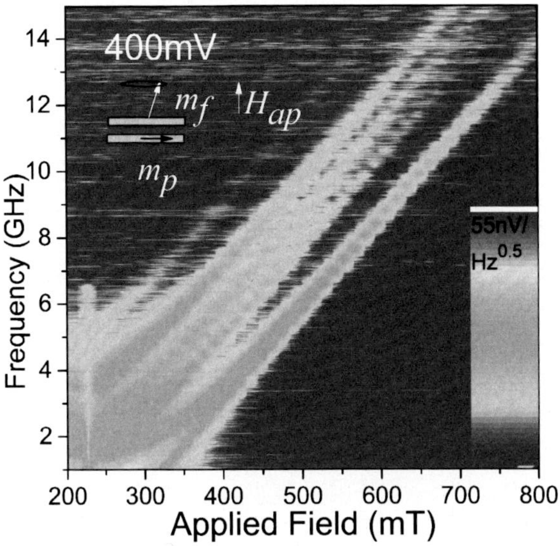

Figure 6.14 Magnetic noise spectra from magnetic tunnel junctions showing multiple modes.

involve oscillations at the end of the bit and may play an important role in high-speed switching of MRAM [30].

6.4.3 *Spin Transfer Devices*

Slonczewski [12] realized that the LLG equation would have to be modified in magnetic devices since the electron currents transport angular momentum from one layer to another:

$$\frac{d\vec{\mathrm{m}}_\mathrm{f}}{dt} = -\mu_0 \frac{|\gamma|}{1+\alpha^2}\vec{\mathrm{m}}_\mathrm{f} \times \vec{\mathrm{H}}_\mathrm{eff} - \alpha \frac{|\gamma|}{1+\alpha^2}\vec{\mathrm{m}}_\mathrm{f} \times (\vec{\mathrm{m}}_\mathrm{f} \times \vec{\mathrm{H}}_\mathrm{eff}) + \frac{g(\theta_\mathrm{m})\mu_\mathrm{e} I}{e\mathbf{M}_\mathrm{s} V}\vec{\mathrm{m}}_\mathrm{f} \times (\vec{\mathrm{m}}_\mathrm{f} \times \vec{\mathrm{m}}_\mathrm{p}) \quad (6.12)$$

The new term accounts for the angular momentum transferred to a free layer moment $\vec{\mathrm{m}}_\mathrm{f}$ by a current I from a layer with polarization direction $\vec{\mathrm{m}}_\mathrm{p}$ (which typically is the fixed or pinned layer shown in Fig. 6.8). The magnitude of the spin transfer term normalized to the precessional term is approximately $\frac{g(\theta_\mathrm{m})\mu_\mathrm{e} I}{\mu_0 \gamma e \mathbf{M}_\mathrm{s}^2 V} \approx 0.03$, where we have assumed that $\mathbf{H}_\mathrm{eff} \approx \mathbf{M}_\mathrm{s}$, an applied current of 1 mA, and a 100 nm device size. Since the spin transfer term is small, on the order of the damping term, it can also be viewed as a small perturbation to the precessional term. The equation of motion with the spin torque term is referred to as the Landau–Lifshitz–Gilbert–Slonczewski (LLGS) equation and it has been extensively studied both numerically [31] and analytically [32].

Depending on the relative orientation of the polarization direction to the effective field, the spin transfer term can either add energy or remove energy from the magnetic system. To understand how these terms affect the magnetization dynamics it is useful to look at a simple high-symmetry configuration. In the case of a perpendicular applied field and polarization the LLGS equation can be rewritten as

$$\frac{d\vec{\mathrm{m}}_{\mathrm{f}}}{dt} = -\mu_0 \frac{|\gamma|}{1+\alpha^2} \vec{\mathrm{m}}_{\mathrm{f}} \times \vec{\mathrm{H}}_{\mathrm{eff}} - \left(\alpha \frac{|\gamma|\, \mu_0 H_{\mathrm{eff}}}{1+\alpha^2} - \frac{g(\theta_m)\mu_{\mathrm{e}} I}{e\mathbf{M}_{\mathrm{s}} V} \right) \vec{\mathrm{m}}_{\mathrm{f}} \times (\vec{\mathrm{m}}_{\mathrm{f}} \times \hat{z}) \tag{6.13}$$

We can see that the spin torque term opposes the damping term if I is positive and adds to the damping if I is negative. When the current becomes greater than a critical current, the effective damping becomes negative, and the system becomes unstable and moves toward orbits of higher energy. Depending on the device configuration, the magnetization may switch to another stable state [33] or may be driven into continuous precession [34–36]. These effects have led to the development of spin transfer torque (STT) MRAM and spin transfer nano-oscillators.

The critical current is given by setting the effective damping $\alpha_{\mathrm{eff}} = \alpha \frac{|\gamma|\mu_0 \mathbf{H}_{\mathrm{eff}}}{1+\alpha^2} - \frac{g_S(\theta_{\mathrm{m}})\mu_{\mathrm{e}} I}{e\mathbf{M}_{\mathrm{s}} V}$ to 0:

$$I_{\mathrm{c}} = \alpha \frac{|\gamma|\, \mu_0 \mathbf{H}_{\mathrm{eff}}}{1+\alpha^2} \frac{e\mathbf{M}_{\mathrm{s}} V}{g_S(\theta_{\mathrm{m}})\mu_{\mathrm{e}}} \tag{6.14}$$

To obtain a small critical current to set the free layer in motion, devices need free layers with small damping constants, low saturation magnetizations, small volumes, and high polarizations. Reducing $\mathbf{H}_{\mathrm{eff}}$ is in general not feasible since a large $\mathbf{H}_{\mathrm{eff}}$ is required to maintain a high operation frequency. Conversely, when designing a fixed polarizer layer, we require a large magnetization, large damping, and a large effective field so that the current is not sufficient to get the polarizer precessing. The above discussion

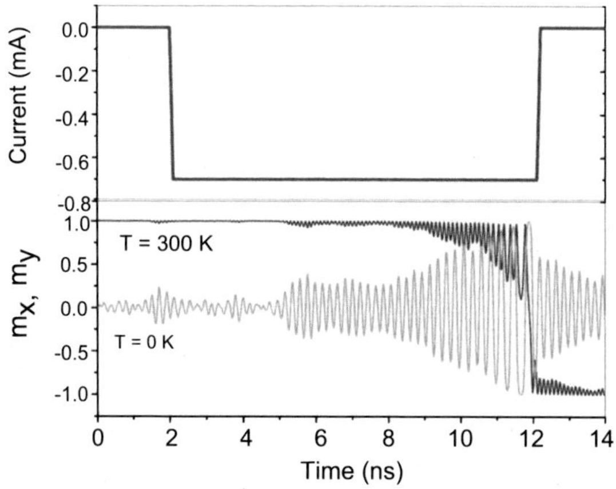

Figure 6.15 Numerical simulation of the LLGS equation for a 50 nm × 150 nm STT-RAM bit showing switching at 300 K but not at 0 K.

applies strictly to the case when everything is symmetric about the z axis. In general, the effective damping will not be zero everywhere on the orbit. Stable orbits are obtained by having the integrated damping and spin torque terms balance on the orbit.

An example of numerical simulations of a 50 nm × 150 nm STT-RAM bit is shown in Fig. 6.15. When current is applied in the positive direction the effective damping goes negative, and the thermal fluctuations grow until the magnetization switches polarity. In the new configuration, the STT changes sign and adds to the damping, causing the moment to settle into a new state. In this simulation, the current is slightly below the $T = 0$ critical current and the bit only switches with the aid of thermal fluctuations.

Thermal fluctuations can therefore aid or hinder switching of these memory cells. These fluctuations can lead to write errors, as shown in Fig. 6.16. The write error rate is the probability that the bit will not switch when a write pulse is applied. Ideally, the write error rate goes to zero above a given threshold. However, as seen in Fig. 6.16, the write error rate can drop off slowly and never reach zero. It is essential that the fluctuations and switching mechanism be understood and optimized to reduce these write errors.

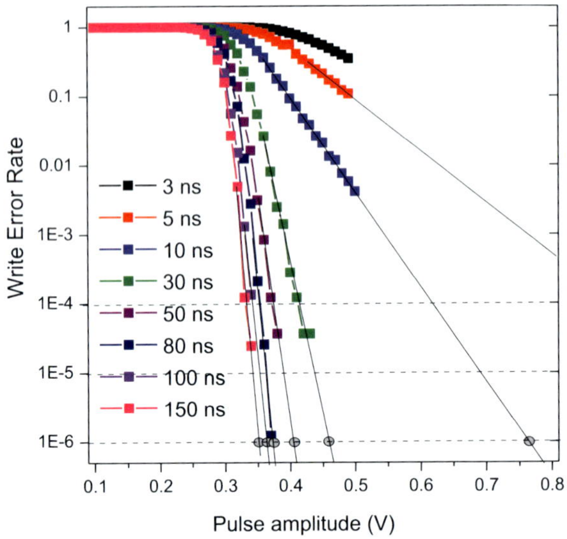

Figure 6.16 Write error rates as a function of write pulse amplitude for a 50 × 150 nm spin torque transfer–random access memory bit for different pulse durations.

6.5 Quantum Effects

As the magnet cluster size decreases quantum effects become apparent and discrete transitions are seen between quantized energy levels. Molecular nanomagnets, whose quantum behavior has been extensively studied, are magnetic clusters involving strongly exchange coupled local spins that have a precise and reproducible structure. Magnetic resonance spectra of Fe-8, a spin-10 molecular magnet, are shown in Fig. 6.17. Here, the *y* axis displays the percentage decrease in moment along the field direction when 141 GHz microwaves are applied. When the applied field is such that a particular transition is in resonance, the system will absorb a photon and the moment will decrease by $\hbar$. The resonant absorption peaks clearly display the transitions between neighboring quantum levels and the peak amplitudes explicitly show the thermal occupation probabilities. The broad magnetic resonance peak due to the presence of an anisotropy energy, which leads to orbits with differing frequencies, is broken up into discrete quantum

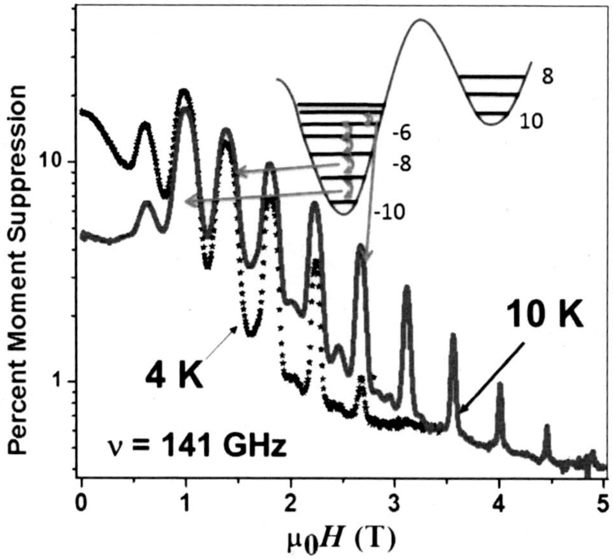

Figure 6.17 Electron paramagnetic resonance spectra of Fe-8 at 4 K and 10 K. Fe-8, at low temperatures, is a *spin 10 molecular magnet with 21 energy levels (designated as* $-10, -9, -8 \ldots 10$*) with an anisotropy energy of* U_{an}/k_B $= 22$ K. Plotted is the percent suppression of the moment along the applied field axis when irradiated with 147 GHz microwaves. The peaks correspond to resonant excitation between quantized energy states as shown in the schematic energy diagram [37].

transitions. At a low temperature of 4 K, only the low-energy states are occupied, while at higher temperatures, 10 K, states up to the energy barrier are occupied. Some of the nanoparticles, whose spectra are shown in Fig. 6.8, may also show quantum effects. However, these effects are masked due to the inability to measure single nanostructures.

6.6 Summary and Outlook

Magnetization dynamics in magnetic nanoparticles or nanodevices are complicated due to the presence of thermal fluctuations. Thermal fluctuations lead to superparamagnetic behavior that may be beneficial applications for contrast or diagnostic agents.

Fluctuations can also lead to instability and unpredictability in nanodevice performance. Understanding the detailed dynamics in magnetic nanostructures is an unsolved problem due to the difficulty of measuring a single nanostructure with GHz bandwidth and nanometer resolution. Ensemble measurements are not sufficient to fully characterize nanoparticle dynamics. For instance, with an ensemble of slightly different nanostructures, ensemble measurements will wash out quantum effects that must come into play as the moments approach 10 μ_B. Understanding and measuring the dynamics of magnetic nanoparticles and nanodevices remains an important unsolved problem that is of both fundamental interest and technical importance.

References

1. J. Akerman, Applied physics. Toward a universal memory, *Science*, **308**(5721), 508–510 (2005).
2. W. J. Gallagher and S. S. P. Parkin, Preface, *IBM J. Res. Dev.*, **50**(1), 3–4 (2006).
3. Q. A. Pankhurst et al., Applications of magnetic nanoparticles in biomedicine, *J. Phys. D: Appl. Phys.*, **36**(13), R167–R181 (2003).
4. J. Lodhia et al., Development and use of iron oxide nanoparticles (part 1): synthesis of iron oxide nanoparticles for MRI, *Biomed. Imaging. Interv. J.*, **6**(2), e12 (2010).
5. G. Mandarano et al., Development and use of iron oxide nanoparticles (Part 2): the application of iron oxide contrast agents in MRI, *Biomed. Imaging Interv. J.*, **6**(2), e13 (2010).
6. Q. A. Pankhurst et al., Progress in applications of magnetic nanoparticles in biomedicine, *J. Phys. D: Appl. Phys.*, **42**(22), 224001 (2009).
7. N. D. Chasteen and P. Harrison, Mineralization in ferritin: an efficient means of iron storage, *J. Struc. Biol.*, **126**, 182–194 (1999).
8. C. Quintana, J. M. Cowley, and C. Marhicc, Electron nanodiffraction and high-resolution electron microscopy studies of the structure and composition of physiological and pathological ferritin, *J. Struct. Biol.*, **147**(2004), 166–178 (2004).
9. W. T. Coffey and Y. P. Kalmykov, Thermal fluctuations of magnetic nanoparticles: fifty years after Brown, *J. Appl. Phys.*, **112**(12), 121301 (2012).

10. N. F. Mott, The electrical conductivity of transition metals, *Proc. R. Soc.*, **153, 156**, 699 (1936).
11. E. Yu. Tsymbal and D. G. Pettifor, Effects of band structure and spin-independent disorder on conductivity and giant magnetoresistance in Co/Cu and Fe/Cr multilayers, *Phys. Rev B*, **54**, 15314 (1996).
12. J. C. Slonczewski, Current-driven excitation of magnetic multilayers, *J. Magn. Magn. Mater.*, **159**, L1–L7 (1996).
13. J. C. Slonczewski, Currents and torques in metallic magnetic multilayers, *J. Magn. Magn. Mater.*, **247**, 324–338 (2002).
14. L. Berger, Emission of spin waves by a magnetic multilayer traversed by a current, *Phys. Rev. B*, **54**, 9353 (1996).
15. A. H. Morrish, *The Physical Principles of Magnetism*, Wiley-IEEE Press, 2001.
16. G. Bertotti et al., *Nonlinear Magnetization Dynamics in Nanosystems*, Elsevier Science and Technology, 2008.
17. R. D. McMichael et al., Ferromagnetic resonance mode interactions in periodically perturbed films, *J. Appl. Phys.*, **91**, 8647–8649 (2002).
18. V. Kambersky, On ferromagnetic resonance damping in metals, *Czech. J. Phys. B*, **26**, 1366 (1976).
19. W. F. Brown, Jr., Thermal fluctuations of a single-domain particle, *Phys. Rev.*, **130**, 1677 (1963).
20. K. Jacobs, *Stochastic Processes for Physicists: Understanding Noisy Systems*, Cambridge University Press, 2010.
21. I. Mayergoyz et al., Landau–Lifshitz magnetization dynamics driven by a random jump-noise process (invited), *J. Appl. Phys.*, **109**(7), 07D312 (2011).
22. N. Smith and P. Arnett, White-noise magnetization fluctuations in magnetoresistive heads, *Appl. Phys. Lett.*, **78**(10), 1448 (2001).
23. N. Smith, Modeling of thermal magnetization fluctuations in thin-film magnetic devices, *J. Appl. Phys.*, **90**(11), 5768 (2001).
24. H. Li et al., Determination of anisotropy constants of protein encapsulated iron oxide nanoparticles by electron magnetic resonance, *J. Magn. Magn. Mater.*, **321**, 175–180 (2009).
25. R. J. Usselman et al., Two-component magnetic structure of iron oxide nanoparticles mineralized in Listeria innocua protein cages, *J. Appl. Phys.*, **107** (2010).
26. R. J. Usselman et al., Temperature dependence of electron magnetic resonance spectra of iron oxide nanoparticles mineralized in Listeria innocua protein cages, *J. Appl. Phys.*, **112**, 084701 (2012).

27. E. de Biasi et al., Size and anisotropy determination by ferromagnetic resonance in dispersed magnetic nanoparticle systems, *J. Magn. Magn. Mater.*, **262**, 235–241 (2003).
28. S. E. Russek et al., Spin dynamics in confined magnetic structures II, in *High Speed Switching and Rotational Dynamics in Small Magnetic Thin Film Devices*, B. Hillebrands and K.Ounadjela, Eds. Berlin: Springer-Verlag, 2003.
29. N. A. Stutzke, S. L. Burkett, and S. E. Russek, High-frequency noise measurements in spin-valve devices, *J. Vac. Sci. Technol. A*, **21**(4), 1167–1171 (2003).
30. R. Heindl et al., Size dependence of intrinsic spin transfer switching current density in elliptical spin valves, *Appl. Phys. Lett.*, **92** (2008).
31. J. Z. Sun, Spin-current interaction with a monodomain magnetic body: a model study, *Phys. Rev. B*, **62**, 570 (2000).
32. G. Bertotti et al., Magnetization switching and microwave oscillations in nanomagnets driven by spin-polarized currents, *Phys. Rev. Lett.* **94**, 127206 (2005)
33. J. A. Katine et al., Current-driven magnetization reversal and spin-wave excitations in Co/Cu/Co pillars, *Phys. Rev. Lett.*, **84**(14), 3149–3152 (2000).
34. M. Tsoi et al., Generation and detection of phase-coherent current-driven magnons in magnetic multilayers, *Nature*, **406**, 46–48 (2000).
35. S. I. Kiselev et al., Microwave oscillations of a nanomagnet driven by a spin-polarized current, *Nature*, **425**, 380 (2003).
36. W. H. Rippard et al., Direct-current induced dynamics in $Co_{90}Fe_{10}$/$Ni_{80}Fe_{20}$ point contacts, *Phys. Rev. Lett.*, **92**, 027201 (2004).
37. B. Cage et al., Resonant microwave power absorption and relaxation of the energy levels of the molecular nanomagnet Fe-8 using superconducting quantum interference device-based magnetometry, *Appl. Phys. Lett.*, **87** (2005).
38. R. Heindl et al., Validity of the thermal activation model for spin-transfer torque switching in magnetic tunnel junctions, *J. Appl. Phys.*, **109** (2011a).
39. R. Heindl et al., Physical limitations to efficient high-speed spin-torque switching in magnetic tunnel junctions, *Phys. Rev. B*, **83** (2011b).
40. R. O. O'Handley, *Modern Magnetic Materials: Principles and Applications*, New York: John Wiley and Sons, 2000.
41. Yu. L. Raikher and V. I. Stepanov, The effect of thermal fluctuations on the FMR line shape in dispersed ferromagnets, *Sov. Phys. JETP*, **75** (1992).

42. D. C. Ralph and M. D. Stiles, Spin transfer torques, *J. Magn. Magn. Mater.*, **320**, 1190–1216 (2008).
43. T. J. Silva and W. H. Rippard, Developments in nano-oscillators based upon spin-transfer point-contact devices, *J. Magn. Magn. Mater.*, **320**, 1260–1271 (2008).
44. J. Z. Sun, Spin angular momentum transfer in current perpendicular nanomagnetic junctions, *IBM J. Res. Dev.*, **50**, 81–100 (2006).
45. S. Yuasa et al., Giant room-temperature magnetoresistance in single-crystal Fe/MgO/Fe magnetic tunnel junctions, *Nat Mater.*, **3**, 868 (2004).

Chapter 7

Nonlinear Optics in Magnetic Nanostructures

Tatiana Murzina and Oleg Aktsipetrov

Physics Department, M.V. Lomonosov Moscow State University, GSP-1, Leninskie Gory, Moscow 119991, Russia

murzina@mail.ru

This chapter makes a review of some of the aspects of the nonlinear optics of magnetic nanostructures. We intend to demonstrate the unique sensitivity of second-order nonlinear optical effects, not only to the structure, morphology, and composition of the nanostructures, but also to the magnetic state of nanostructures.

7.1 Introduction

Active studies in the field of nonlinear optics of surfaces, interfaces, and thin films trace back to the beginning of the 1980s and was caused by both the progress in laser sources and the new achievements in the technology for the fabrication of such nanostructures. From the other side, research activity in this field was stimulated by novel effects and exciting application of nanostructures. Magnetic nanostructures have a special place

Handbook of Nanomagnetism: Applications and Tools
Edited by Rosa A. Lukaszew

ISBN 978-981-4613-04-0 (Hardcover), 978-981-4613-05-7 (eBook)
www.panstanford.com

among different types of nanostructures as they reveal such effects as giant magnetoresistance (GMR).

Thus two aspects of the research were of extremely high interest: first, the development of the nonlinear magneto-optics of magnetic nanostructures and the search of possible new effects caused by the nanometer size of these structures and, second, the development of effective experimental methods that may supply additional information on the magnetic state of the structures that can be hardly attained by commonly used techniques. And here the most impressive results were demonstrated using the second-harmonic generation (SHG) method when exploiting its unprecedentedly high sensitivity.

It arises from the symmetry selection rules for the second-order susceptibility. In the dipole approximation, the SHG is strongly forbidden in the bulk of centrosymmetric materials but is allowed for structures with a lack of inversion symmetry, that is, at surfaces, interfaces etc., which is extremely important for the studies of nanostructured materials.

7.2 Historical Aspects of Nonlinear Magneto-Optics

Conventional magneto-optics, with its more than a century-long history, remains one of the most important experimental methods in studies of magnetism. Recently, significant attention has been directed toward the nonlinear optical analog, that is, nonlinear magneto-optics. Nonlinear magneto-optics deals with the influence of an external magnetic field on the nonlinear optical effects.

Nonlinear magneto-optics traces back to the mid-1980s, when the dependence of the SHG intensity and second-harmonic (SH) wave polarization on magnetic fields was observed [1, 2]. Both the nonlinear magneto-optical Kerr effect (NOMOKE) (in reflection) and the nonlinear optical Faraday effect (in transmission) for the magnetization-induced second-harmonic generation (MSHG) were observed in thin magnetic garnet films. It is important that our 1988 experiments were initiated by the 1985 theoretical paper by Akhmediev, Zvezdin and coworkers, who predicted nonlinear magneto-optical effects in the bulk of magnetic materials [3]. As for the following history of nonlinear magneto-optics, Shen and coworkers considered the theory of the surface MSHG in their 1989 paper [4] and in 1991, Kirschner and coworkers studied the MSHG in reflection from the surface of an Fe single crystal [5]. They studied the influence of surface magnetization on the optical SHG in reflection at an Fe(110) surface in high-vacuum conditions. For the atomically clean surface, a magnetic effect on SHG intensity of approximately 0.25 was measured. This magnetic contrast of the SHG intensity was shown to decay exponentially with the surface contamination by the residual gas in a vacuum chamber. This key feature of Kirschner's experiment demonstrates the high surface sensitivity of the MSHG.

Thanks to this surface and interface sensitivity, the MSHG probe is potentially perspective for the studies of nano- and surface magnetism in magnetic nanostructures. First of all, planar layered magnetic structures—magnetic quantum wells and superlattices—were intensively studied by the MSHG [6–8]. Another class of magnetic nanostructures, magnetic nanogranular films, was investigated as well [9–12]. Both types of magnetic nanostructures have received considerable attention recently because of new magnetic phenomena observed in these systems: the oscillatory coupling through the nonmagnetic spacers, spin-dependent electron scattering and tunneling, and the GMR effect.

The idea of our experiments was the simultaneous experimental characterization of magnetic nanogranular films by means of GMR and the MSHG, with a goal to find correlation between these two phenomena [9, 10]. Thus, the magnetization-induced SHG was experimentally studied in Co_xAg_{1-x} and $(CoFe)_x(Al_2O_3)_{1-x}$ granular films which exhibit the GMR effect. Nonmonotonous dependences of the magnetoresistance and magnetic contrast of the SHG intensity as a function of the volume fraction of the magnetic component x are found. The maximum of MSHG magnetic contrast is attained at the same value of x as that providing the maximum of the GMR coefficient. This overlap of extrema of GMR and MSHG parameters can be interpreted as a correlation between magnetotransport and nonlinear magneto-optical effects.

Nonlinear magneto-optics would be a poor area if magnetic effects were observed only for the SHG. Meanwhile, a magnetization-induced third-harmonic generation (THG) was found in magnetic nanostructures [9] and magnetophotonic crystals (MPCs) [13]. Details of nonlinear magneto-optical effects in magnetic nanostructures can be found in Ref. [14]. In this paper, the results of our recent experimental studies of magnetization induced second- and third-order nonlinear optical effects in magnetic nanostructures are surveyed. Magnetization induced variations of the intensity, polarization state, and relative phase of SH waves are studied in magnetic nanogranular films, self-assembling films with magnetic garnet nanoparticles, thin magnetic metal films, and rare-earth-containing LB films. The nonlinear magneto-optical Kerr effect (NOMOKE) in the SHG from thin magnetic and granular films exceeds the linear magneto-optical Kerr effect (MOKE) by at least an order of magnitude. Magnetization-induced optical THG is observed in magnetic thin metal films and nanogranular films. The NOMOKE in the THG from these magnetic nanostructures appears to be of the same order of magnitude as the second-order NOMOKE in the SHG. The NOMOKE magnetic contrast in intensity of magnetization-induced THG is up to $\sim$0.1 in Co_xAg_{1-x} nanogranular films. For THG waves, the magnetization induced rotation angle of polarization is up to 10° in thin Fe(110) films and the relative phase shift is up to 70° in thin Co films. The studies of magnetization-induced quadratic and cubic nonlinear optical effects show the interconnection between the magnetic, structural, and magneto-optical properties of magnetic nanomaterials.

A novel class of artificial magneto-optical materials has appeared recently. In 1999, Inoue and coworkers introduced MPCs, that is, periodic magnetic structures that exhibit both magneto-optical properties and photonic band gap [15]. For the last decade, we have developed nonlinear magneto-optics of MPCs in collaboration with Inoue. Magnetization-induced second- [16] and third-harmonic (TH) [13] generation were observed and systematically studied [17].

In this chapter we discuss (1) the phenomenological description of the magnetization-induced SHG and THG, (2) magnetization-induced anisotropy of the nonlinear optical response in the simplest case of an isotropic surface, (3) the MSHG studied multilayer structures (4) the NOMOKE in magnetoresistive granular films, and

(5) the NOMOKE in inhomogeneous arrays of yttrium-iron-garnet (YIG) nanoparticles

7.3 Phenomenological Description of Nonlinear Magneto-Optical Effects

Nonlinear magneto-optics describes the nonlinear optical effects that are governed by the magnetic properties of a medium. As was noted in the introductory part, these are the nonlinear optical analogues of the well-known magneto-optical Faraday and Kerr effects. Here we will mostly note them as the effects induced by the *magnetization* instead of the external magnetic field. This is the better choice as the MSHG effects can be detected in the case of ferromagnetic materials due to their remanent magnetization and in the absence of an external magnetic field.

The description of the magneto-optical effects in a magnetic medium is based on the wave equation

$$\nabla \times \nabla \times \mathbf{E}_{\omega} + \frac{\varepsilon(\omega)}{c^2}\frac{\partial^2}{\partial t^2}\mathbf{E}_{\omega} = 0, \tag{7.1}$$

where $\mathbf{E}_{\omega}$ is the electromagnetic field, c is the speed of light, and $\varepsilon(\omega)$ is the permittivity tensor of the medium that determines the magneto-optical effects and for the case of the magnetization along the z axis has the form

$$\hat{\varepsilon} = \begin{pmatrix} \varepsilon & -i\varepsilon Q & 0 \\ i\varepsilon Q & \varepsilon & 0 \\ 0 & 0 & \varepsilon \end{pmatrix}, \quad Q \propto \mathbf{M}. \tag{7.2}$$

Here **M** is the magnetization and the case of a spatially homogeneous medium is considered.

On the contrary, in the nonlinear optical case the process of the second- or third-harmonic generation is described by an inhomogeneous wave equation

$$\nabla \times \nabla \times \mathbf{E}_{2\omega,3\omega} + \frac{\varepsilon(2\omega, 3\omega)}{c^2}\frac{\partial^2}{\partial t^2}\mathbf{E}_{2\omega,3\omega} = -\frac{1}{\varepsilon_0 c^2}\frac{\partial^2}{\partial t^2}\mathbf{P}_{2\omega,3\omega} \tag{7.3}$$

where $\mathbf{E}_{2\omega,3\omega}$ is the SHG or THG field, $\varepsilon(2\omega, 3\omega)$ is the permittivity tensor at the corresponding wavelengths, and $\mathbf{P}_{2\omega,3\omega}$ is the nonlinear polarization induced in a nonlinear medium by the light field. In the simplest electric dipole approximation $\mathbf{P}_{2\omega} = \hat{\chi}^{(2)D} : \mathbf{E}_\omega \mathbf{E}_\omega$, $\mathbf{P}_{3\omega} = \hat{\chi}^{(3)D} : \mathbf{E}_\omega \mathbf{E}_\omega \mathbf{E}_\omega$, where $\hat{\chi}^{(2)D}$ and $\hat{\chi}^{(3)D}$ are the second- and third-order nonlinear susceptibility tensors. An important property of the SHG is that it vanishes in the bulk of centrosymmetric media due to symmetry considerations [18]. Meanwhile, any type of symmetry breaking, including the cutoff of the crystalline structure at surfaces and interfaces, brings about the appearance of the second-order dipole susceptibility. This property leads to a well-known extremely high surface sensitivity of the SHG effect, which is of particular importance for studies of nanostructures.

At the same time, the THG is allowed for media of any symmetry class and thus can be applied for getting supplementary information as compared to SHG.

It has been shown in Ref. [4] that for a particular case of magnetized medium two types of the susceptibility components linear in the magnetization can exist, which are *odd* and *even* in **M**. The first one, $\hat{\chi}^{(2)/(3)\text{odd}}(\mathbf{M})$, changes its sign under the reversal of magnetization, $-\hat{\chi}^{(2)/(3)\text{odd}}(\mathbf{M}) = \hat{\chi}^{(2)/(3)\text{odd}}(-\mathbf{M})$, while the second one, $\hat{\chi}^{(2)/(3)\text{even}}(\mathbf{M})$, does not change. These components are the sources of the magnetization-induced SH and TH waves as they contribute to the nonlinear polarization:

$$\begin{aligned} \mathbf{P}^{\text{NL}}_{2\omega}(\mathbf{M}) &= \left[\hat{\chi}^{(2)\text{even}} + \hat{\chi}^{(2)\text{odd}}(M)\right] : \mathbf{E}_\omega \mathbf{E}_\omega \\ \mathbf{P}^{\text{NL}}_{3\omega}(\mathbf{M}) &= \left[\hat{\chi}^{(3)\text{even}} + \hat{\chi}^{(3)\text{odd}}(M)\right] \vdots\ \mathbf{E}_\omega \mathbf{E}_\omega \mathbf{E}_\omega \end{aligned} \tag{7.4}$$

Let us consider briefly the case of the SHG from an isotropic metal surface (a similar description is easily extended for the THG). The symmetry of the $\hat{\chi}^{(2)}$ tensor is determined by the symmetry of a particular medium, so that the invariance over the symmetry operation should be conserved:

$$\chi^{(2)}_{ijk} = T_{ii'} T_{jj'} T_{kk'} \chi^{(2)}_{i'j'k'}, \tag{7.5}$$

where $\hat{T}$ is the transformation matrix for each symmetry operation. In a magnetized medium an additional symmetry operation involving the inversion of time appears, thus dividing all the $\hat{\chi}^{(2)}$ components into *odd* and *even* with respect to the magnetization. It can be shown [4] that in a nondissipative medium *odd* in

Table 7.1 Nonzero tensor components of the second-order susceptibility tensor for an isotropic metal surface

Combination of polarizations	$\chi^{\mathbf{even}}(2\omega)$	$\chi^{\mathbf{odd}}(2\omega)$	
		$\mathbf{M} \parallel (OX)$	$\mathbf{M} \parallel (OY)$
p-in, *p*-out	$\chi_{zzz}, \chi_{zxx}, \chi_{xxz}$	0	$\chi_{xxx}, \chi_{xzz}, \chi_{zzx}$
p-in, *s*-out	0	χ_{yxx}, χ_{yzz}	0
s-in, *p*-out	χ_{zyy}	0	χ_{xyy}
s-in, *s*-out	0	χ_{yyy}	0

magnetization susceptibility should be imaginary quantities, while *even* susceptibility components are real, so that there is no interference between them. Besides, the symmetry of *even* in **M** and of nonmagnetic (crystallographic) susceptibility components is the same, so in what follows we will consider only *even* and *odd* components, taking into account that $\hat{\chi}^{\text{even}}(\mathbf{M})$ is mostly determined by the crystallographic SHG contribution.

It is worth noting that in the case of widely studied structures such as ferromagnetic metals or iron-garnet-based films in the visible spectral range optical losses cannot be neglected, and thus the interference between the SH fields driven by $\hat{\chi}^{\text{even}}(\mathbf{M})$ and $\hat{\chi}^{\text{odd}}(\mathbf{M})$ is expected.

Table 7.1 summarizes the nonzero even and odd in magnetization $\hat{\chi}^{(2)}$ components for an isotropic surface for different orientations of the magnetization. The coordinate frame is chosen in such a way that the *OZ* axis corresponds to the normal to the surface, while *OX* and *OY* axes lie in the plane of the surface, (ZOX) being the plane of incidence. According to the definition of the MOKE, $\mathbf{M} \parallel (OX)$ corresponds to the longitudinal and $\mathbf{M} \parallel (OY)$ to the transversal MOKE.

Coming back to Eq. 7.3 we see that the effect of the magnetization on harmonics' generation is determined not only by the magnetic part of $\varepsilon(2\omega, 3\omega)$. Additional (as compared to Eq. 7.1) magnetization-induced terms appear in the nonlinear polarization as well. It was noted in Ref. [19] that it is the last contribution that governs the nonlinear magneto-optics and leads to much larger values of the magnetization-induced effects as compared to MOKE.

Let us consider magnetization-induced effects in the SHG response from an isotropic surface. This approximation is valid for the metals like Co, Fe, and Ni that are typically used for making thin ferromagnetic films and nanostructures. As is well known as it stems from Table 7.1, for isotropic surfaces only a p-polarized SH wave can be generated. This is a manifestation of the so-called s-prohibition rule, that is, the forbiddance for the generation of the s-polarized isotropic SH radiation (i.e., independent of the azimuthal orientation of the sample with respect to the plane of incidence) [20, 21].

At the same time, for a magnetized structure and for the longitudinal NOMOKE s-polarized components of $\hat{\chi}^{(2)}$ appear, which bring about magnetization-induced rotation of the SH polarization plane out of the p-polarization. Due to the *odd* character of $\hat{\chi}^{\text{odd}}(\mathbf{M})$ the rotation of the SH polarization plane takes place in the opposite directions as the direction of the magnetization is reversed. Magnetization-induced rotation of the SH polarization plane can be characterized by the Kerr angle of rotation, which can be introduced for the case of the SHG as

$$\Phi_K = a\tan\left(E_{2\omega}^{\text{odd}}(\mathbf{M})\big/E_{2\omega}^{\text{even}}\right) = a\tan\left(E_{2\omega}^{\text{s}}\big/E_{2\omega}^{\text{p}}\right) \tag{7.6}$$

Similar magnetization-induced rotation of the SH polarization plane can be observed for the polar NOMOKE, while the effect is determined by other $\hat{\chi}^{\text{odd}}(\mathbf{M})$ components.

Another situation takes place for the transversal NOMOKE. As is clear from Table 7.1, transversal magnetization does not rotate the polarization plane of the SH wave, while it changes the intensity and phase of the p-polarized SHG instead. The magnetization-induced intensity variations are described by the SHG magnetic contrast as

$$\rho_{2\omega} = \frac{I_{2\omega}(\mathbf{M}) - I_{2\omega}(-\mathbf{M})}{I_{2\omega}(\mathbf{M}) + I_{2\omega}(-\mathbf{M})} \tag{7.7}$$

where $I_{2\omega}(\mathbf{M})$ and $I_{2\omega}(-\mathbf{M})$ are the SHG intensity measured for the opposite directions of the applied magnetic fields. SHG magnetic contrast can be observed only in absorbing magnetic media as the interference between *even* and *odd* in magnetization SH fields takes place. If so, the measured SHG intensity is determined by the square of the vector summation

$$I_{2\omega}(\mathbf{M}) \propto (\mathrm{E}_{2\omega}^{\mathrm{even}} \pm \mathrm{E}_{2\omega}^{\mathrm{odd}}(\mathbf{M}))^2 = (\mathrm{E}_{2\omega}^{\mathrm{even}})^2 \pm 2\mathrm{E}_{2\omega}^{\mathrm{even}}\mathrm{E}_{2\omega}^{\mathrm{odd}}(\mathbf{M})\cos\phi + (\mathrm{E}_{2\omega}^{\mathrm{odd}}(\mathbf{M}))^2 \quad (7.8)$$

where ϕ is the phase shift between *even* and *odd* in magnetization SH fields. If we suppose that $\hat{\chi}^{\mathrm{odd}}(\mathbf{M}) < \hat{\chi}^{\mathrm{even}}(\mathbf{M})$, then

$$\rho_{2\omega} \approx 2\frac{E_{2\omega}^{\mathrm{odd}} \cdot \cos\phi}{E_{2\omega}^{\mathrm{even}}} \propto \frac{\chi^{(2)\mathrm{odd}}}{\chi^{(2)\mathrm{even}}}\cos\phi \quad (7.9)$$

that is, the SHG magnetic contrast is determined by both the ratio of *odd* and *even* in magnetization components of the quadratic susceptibility and by the phase shift ϕ as well. This quantity can be measured experimentally by using the SHG interferometry method applied for a special case of the MSHG [17]. The scheme of this method is shown in Fig. 7.1. An additional SHG reference nonmagnetic sample (e.g., nanometer-thick indium tin oxide [ITO] film) is introduced, so the intensity of the interfering SH fields from the reference and the sample in the far field is measured as a function of the relative distance between the sample and the reference, *d*. Due to the dispersion of air at the fundamental and SHG wavelengths, this dependence is a periodic function of *d* in accordance with the following expression:

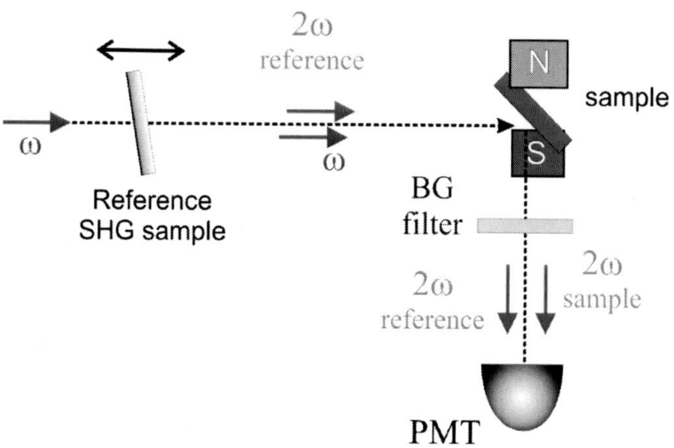

Figure 7.1 Scheme of the SHG interferometry measurements.

$$I_{2\omega} \propto \left|E_{2\omega}^{\text{ref}}\right|^2 + \left|E_{2\omega}^{\text{Sam}}(M)\right|^2 + 2E_{2\omega}^{\text{ref}}E_{2\omega}^{\text{Sam}}(M)\cos \times \left\{(k_{2\omega} - 2k_\omega)\,d - \left(\Phi^{\text{ref}} - \Phi^{\text{Sam}}(M)\right)\right\} \tag{7.10}$$

where $E_{2\omega}^{\text{ref}}$ and $E_{2\omega}^{\text{Sam}}$ are the SH fields generated by the reference and the sample and Φ^{ref} and Φ^{Sam} are their phases. As the phase Φ^{Sam} changes its value as the direction of the magnetization is reversed, the magnetization-induced shift of the interference pattern is observed.

As was shown in Ref. [12], that the combination measurements of the SHG contrast and magnetization-induced phase shift allow us to estimate both $\frac{\chi^{(2)\text{odd}}}{\chi^{(2)\text{even}}}$ and ϕ, which totally characterizes the relative values of the magnetization-induced effects in the SHG.

7.4 Magnetization-Induced Anisotropy of the Nonlinear Optical Response

Let us demonstrate how magnetization can alter the nonlinear response of an isotropic structure. This can be illustrated for the case of in-plane magnetization [22]. As the experimental sample we took an isotropic cobalt film with a thin protective gold layer less than 1 nm thick on top, and the structure was deposited on Si(001) wafer by ion-beam sputtering in a single deposition cycle, as described in Ref. [23]. It was checked that for a demagnetized sample the SHG output was p-polarized and the isotropic, s-polarized SHG component was negligible.

Magnetization-induced anisotropy of the SHG intensity reflected from an isotropic Au/Co/Si structure was studied when rotating the sample together with the permanent magnetic field around the axis normal to the plane of the sample. Figure 7.2 shows the dependencies of the SHG intensity on the azimuthal angle of the sample, Ψ, for the p-in, p-out combination of polarizations of the fundamental and SH waves. Due to the experimental conditions the Ψ angle also determines the azimuthal orientation of the (in-plane) magnetic field, so that $\Psi = 0^\circ$ and $\Psi = 180^\circ$ correspond to the transversal magnetic field. It can be seen that a pronounced anisotropy of the SHG intensity appears contrary to the case of the demagnetized sample.

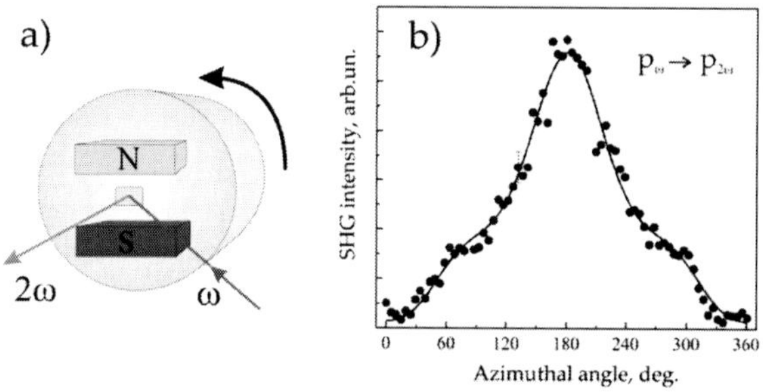

Figure 7.2 (a) Scheme of the experiment on magnetization-induced SHG anisotropy and (b) experimental SHG azimuthal dependence measured in reflection from a Au/Co film for the *p*-in, *p*-out combination of polarizations.

The appearance of the SHG magnetic contrast introduced by (7.7) can be easily seen from the dependencies shown in Fig. 7.2. Evidently, it is caused by the changes in the SHG intensity as the transversal component of the direct current (DC) magnetic field is being reversed, which corresponds to the azimuthal angle values $\Psi = 0^\circ$ and $\Psi = 180^\circ$. The SHG contrast estimated from Fig. 7.2 for the *p*-in, *p*-out combination of polarizations is about 80%. At the same time, SHG interferometry measurements have shown that the magnetization-induced phase shift of the SH from the sample is about $10^\circ - 20^\circ$ and can be neglected in the approximation of the experimental data.

Figure 7.3 shows the experimental dependencies measured for the *s*-polarized SHG and for the *p*- and *s*-polarized fundamental radiation. It can be seen that an anisotropic *s*-polarized SHG appears that is absent for a demagnetized structure and should be referred to as a *purely magnetic* nonlinear optical response.

The presented SHG anisotropic dependencies can be described in terms of the magnetization-induced second-order susceptibility components summarized in Table 7.1. The SHG intensity is determined by

$$I_{2\omega}(\Psi) \propto \left|\mathbf{E}_{2\omega}^{\text{even}}(\Psi) + \mathbf{E}_{2\omega}^{\text{odd}}(\Psi)\right|^2 \propto \left(\chi_{2\omega}^{\text{even}} + e^{i\varphi}\chi_{2\omega}^{\text{odd}}\right)^2 \qquad (7.11)$$

where the contributing $\chi^{(2)}$ components are determined by the particular combinations of polarizations and can be easily obtained from Table 7.1. For the simplest case of the *s*-in, *s*-out SHG

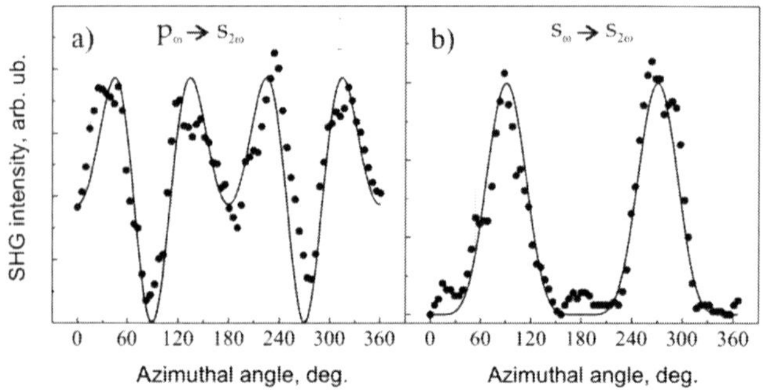

Figure 7.3 SHG azimuthal dependencies measured in reflection from a Co/Au film for the *s*-polarized SHG and for the *p*- and *s*-polarized fundamental radiation.

the azimuthal dependence is given by $I_{2\omega}^{s-s}(\Psi) \propto (a^* \sin(\Psi) + b^* \sin(3\Psi))^2$, where the coefficients a and b are given by proper combinations of the pure magnetic components χ_{yyy} and χ_{xyy}. The results of the corresponding approximation are shown in Figs. 7.2 and 7.3 by solid lines and stay in a good agreement with the experimental data.

It should be noted that we considered here the simplest case of an structurally isotropic magnetic film, while magnetization-induced changes in the SHG anisotropy have been studied in a number of papers [24, 25]; a purely magnetic SHG was also observed, for example, in Ref. [26]. Meanwhile, the described example clearly demonstrates that it is necessary to be quite careful in analyzing the crystallographic symmetry of a structure using the SHG technique, as the huge values of the MSHG effects can bring about strong anisotropy purely induced by remanent magnetization.

7.5 NOMOKE in Magnetoresistive Granular Films

Granular magnetoresistive films are an important and actively studied class of magnetic nanostructures. Apart from extraordinary magnetotransport properties, arrays of magnetic metal nanoparticles may be expected to exhibit unusual optical effects, for example, attributed to the resonant excitation of localized surface plasmons

(LSPs) [27, 28]. On the other hand, as was first demonstrated by Wokaun et al. [28] for the surface-enhanced SHG in silver island films, plasmon excitation can drastically influence the nonlinear optical response of metal nanoparticles due to the resonant enhancement of the local optical field at the fundamental or SH wavelengths. Such a mechanism was shown to result in an increase of the SHG intensity by up to 3 orders of magnitude in silver nanoparticles [29]. To describe this effect local field (LF) factors at the fundamental and SH wavelengths can be introduced so that the second-order polarization of an array of small metal particles is expressed as $P_{2\omega} = L_{2\omega}\chi^{(2)} \vdots L_{\omega}^{2} E_{\omega} E_{\omega}$. Here $\chi^{(2)}$ is the second-order susceptibility of the metal, $L_{2\omega}$, L_{ω} are the LF factors that are anisotropic in the general case, and the symbol $\vdots$ is related to the convolution of the nonlinear susceptibility tensor and the vectors of the fundamental optical field. Thus the resonant increase of the LF factors at the fundamental or SH wavelengths can result in a many-fold increase of the SHG intensity from metal nanopaticles.

Similar considerations initiated the studies of the spectroscopy of the NOMOKE in magnetic granular films of the composition and the corresponding results are presented in this section.

The samples of magnetic nanogranular Co_xAg_{1-x} films were composed by the co-evaporation of Co and Ag from two independent sources by electron-beam deposition on ceramic substrates, as described in detail in Ref. [30]. The structure of the films studied by X-ray diffraction and for the concentration of the magnetic component $x < 0.4$ reveals the existence of Co nanograins of the average diameters from 3 to 6 nm. The films exhibit the GMR effect at room temperature; GMR is characterized qualitatively by the GMR contrast introduced as

$$\rho_{\mathrm{MR}} = -\frac{R(0) - R(\mathbf{H})}{R(0)} \tag{7.12}$$

where $R(0)$ and $R(\mathbf{H})$ are the resistivity measured for the zero and **H** values of the magnetic field. A typical GMR contrast dependence on the magnetic field strength is shown in Fig. 7.4.

The first experiments on the SHG in granular films were performed in Refs. [10, 11] when using a Nd:YAG^{3+} laser operating at 1064 nm as the fundamental radiation. An in-plane magnetic field

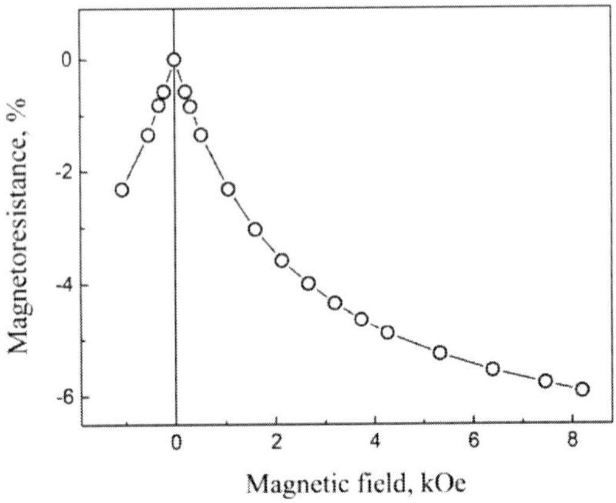

Figure 7.4 GMR contrast vs. magnetic field strength for a $Co_{0.35}Ag_{0.65}$ granular film.

of the strength of 2 kOe was applied to the films using the permanent *Fe-Nd-B* magnets. The dependence of the SHG magnetic contrast on the content of the magnetic component in the films, *x*, was studied and compared with the similar dependence of the GMR coefficient. Figure 7.5 shows the obtained results.

The following features can be noted here. First, GMR coefficient reveals a maximum for the content of cobalt $x \approx 0.3$, that is,

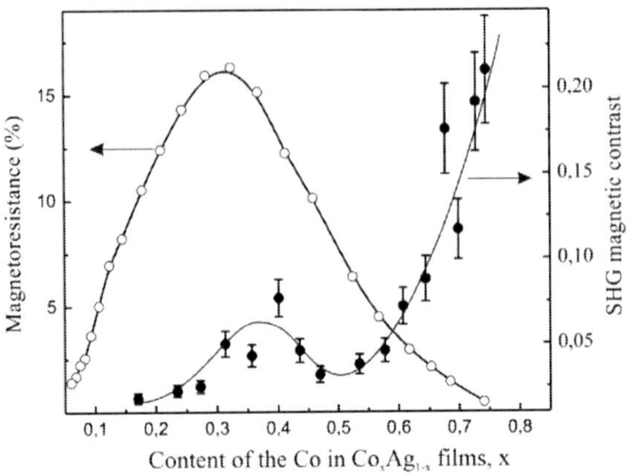

Figure 7.5 GMR coefficient (left scale) and SHG magnetic contrast (right scale) as a function of the content of Co, *x*, in Co_xAg_{1-x} films.

below the percolation threshold; the GMR coefficient reached the value of 16% at room temperature. Second, a nonmonotonous dependence $\rho_{2\omega}(x)$ is seen. An increase of the SHG contrast for $x > 0.5$ corresponds to the formation of the ferromagnetic state as the percolation takes place, the values of $\rho_{2\omega}$ exceed 20% that is typical for surfaces of ferromagnetic metals. An unexpected feature is that an intermediate $\rho_{2\omega}$ maximum was observed for $x \approx 0.3$, which correlates with the GMR maximum. This effect was supposed to be due to the excitation of LSPs in the granular structure [10].

This suggestion required further investigation and first of all spectroscopic SHG studies. Such experiments were made for similar Co_xAg_{1-x} films [14]. The SHG spectroscopy was performed when using an output of a nanosecond OPO laser system as a fundamental radiation, in the wavelength range from 450 nm to 1000 nm, using the signal or the idler beams. The SHG radiation reflected from the granular films at an angle of 45° was spectrally selected by appropriate color filters and a monochromator and detected by a photomultiplier (PMT) and gated electronics. To normalize the SHG signal over the OPO laser fluency and the spectral sensitivity of the optical detection system, an independent reference channel was used with a Z-cut quartz plate as a reference SHG source and a detection system identical to that in the "sample" arm.

To study the spectra of the nonlinear magneto-optical response the dependencies of the SHG intensity on the fundamental wavelength were measured for the opposite directions of the magnetic field and for the *p*-in, *p*-out combination of polarizations and the SHG contrast $\rho_{2\omega}$ was calculated. Such spectra for the granular films with $x = 0.1$, 0.17, and 0.35 are shown in Fig. 7.6. It can be seen that the spectra reveal a resonant-like behavior, the spectral position of the peak as well as its amplitude being dependent on the content of the magnetic component.

Figure 7.7 shows the spectra of $\rho_{2\omega}$ for a $Co_{0.35}Ag_{0.65}$ granular film along with the spectral dependence of the SHG intensity measured in the absence of a magnetic field. It demonstrates a complicated resonant structure: a strong and sharp peak is centered in the vicinity of 4 eV, and a less intensive broad band covers the spectral range from 4.2 eV to 4.8 eV. The spectral position of the

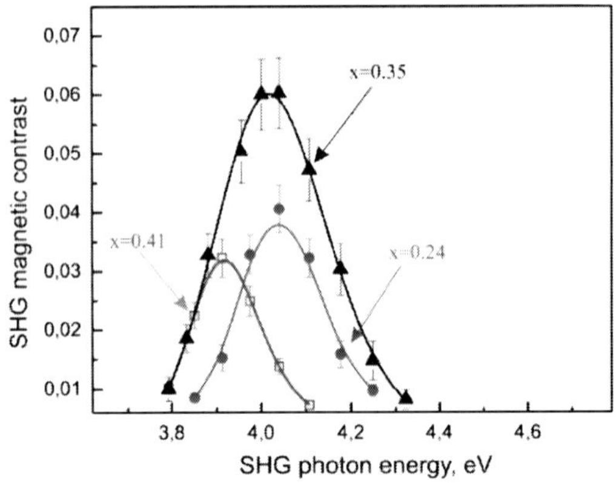

Figure 7.6 Spectra of the SHG magnetic contrast for Co_xAg_{1-x} films, $x = 0.24$, 0.35, and 0.41.

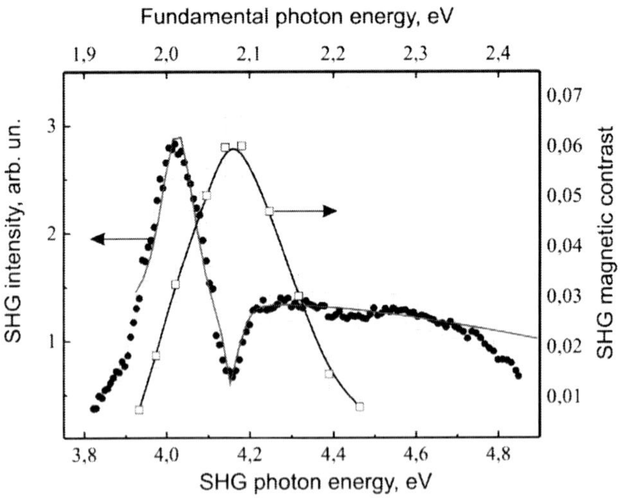

Figure 7.7 Spectra of the SHG intensity in the absence of a magnetic field (left scale) and of the SHG magnetic contrast (right scale) for a $Co_{0.35}Ag_{0.65}$ granular film.

$\rho_{2\omega}$ maximum corresponds to the minimum of the nonmagnetic SHG. Taking into account that $\rho_{2\omega} \propto \chi^{(2)\text{odd}}/\chi^{(2)\text{even}}$ this means evidently that $\chi^{(2)\text{odd}}$ and $\chi^{(2)\text{even}}$ can have different spectral dependencies. Besides, as no such resonance features exist for bulky cobalt and silver, the observed spectral dependence thus should be attributed to the nanograin structure of the films.

It was supposed that the LSP excitation in the granular magnetic structure may be the key process that results in such a resonant-like spectral behavior of both the nonmagnetic and the magnetization-induced SHG. It was shown in Ref. [14] that for Co-Ag granular films and for the spectral range of two-photon energy from 3.8 eV to 4.5 eV the dielectric constant of silver is positive, and thus the condition for the excitation of LSP in cobalt granules is fulfilled. The split of the LSP modes in the spectral doublet evident in the case of the nonmagnetic SHG can be caused by the deviation of the shape of Co nanoparticles from spherical. The resonant character of the observed effect is also proven by the dependence of the resonant wavelength λ_{res} on Co content: increase of x results in a red shift of the LSP resonant wavelengths due to an increase of the interparticle dipole–dipole interaction, which is apparently seen in the set of the spectra of the SHG magnetic contrast in Fig. 7.6

The experimental spectra are approximated within the model that considers the anisotropic LF factors and the NOMOKE under the resonant excitation of LSP in an array of small metal spheroids embedded in a dielectric matrix. According to Ref. [29], the spectral dependence of the anisotropic LF factor, $L_{\mathrm{a}}(\lambda)$, is given by

$$L_{\mathrm{a}}(\lambda) = \frac{\varepsilon_{\mathrm{d}}(\lambda)}{\varepsilon_{\mathrm{d}}(\lambda) + [\varepsilon_{\mathrm{m}}(\lambda) - \varepsilon_{\mathrm{d}}(\lambda)]\,(N_\alpha - \beta x)} \tag{7.13}$$

where $\varepsilon_{\mathrm{d}}(\lambda)$ and $\varepsilon_{\mathrm{m}}(\lambda)$ are the complex dielectric constants of the matrix and of the metal spheroids, λ is the wavelength of light, β is the Lorenz field factor and x is the filling factor, that is, the relative fraction of the metal in a composite material; N_α is an anisotropic depolarization factor of spheroids; the subscript $\alpha = \|$, $\perp$ denotes the tangential and normal orientation of principle semiaxes of spheroids with respect to the sample surface. The resonant wavelength of the LF factor, λ_{res}, corresponds to setting the real part of the denominator in Eq. 7.13 to zero and should depend on the orientation of spheroids within the sample and with respect to the polarization of the laser beam.

The solid line in Fig. 7.7 shows the result of such an approximation, which stays in good agreement with the experimental data and gives the ratio of the semiaxes of the cobalt spheroids in a Ag matrix $a/b = 0.5 \pm 0.1$ [14].

In a similar way the spectral behavior of the SHG magnetic contrast can be explained. Again, granular films are assumed to consist of small Co spheroids oriented in the similar way, with semiaxes a and b parallel and perpendicular to the film surface, respectively. We suppose that two LSPs are excited at the SHG wavelength that correspond to the two axes of the spheroids and result in the appearance of complex LF factors $L_{\|}(2\omega) = C_{\|}e^{-i\varphi_{\|}}$ and $L_{\perp}(2\omega) = C_{\perp}e^{-i\varphi_{\perp}}$, where $C_{\|}$, $C_{\perp}$ and $\varphi_{\|}$ $\varphi_{\perp}$ are the real amplitudes and phases of the anisotropic LF factor, respectively. As the symmetry of *even* and *odd* in magnetization $\chi^{(2)}$ components is different (see Table 7.1), the main contribution to the enhancement of the magnetization-induced SHG component is given by $L_{\|}(2\omega)$, while $L_{\perp}(2\omega)$ determines mostly the resonant enhancement of the nonmagnetic (*even* in **M**) SHG. In thus assumption and using the Eq. 7.9 for $\rho_{2\omega}$ one can obtain the following formula for the spectral dependence of the SHG magnetic contrast:

$$\rho_{2\omega} \approx \frac{C_{\|}\chi^{(2)\text{odd}}_{\|jk}}{C_{\perp}\chi^{(2)\text{even}}_{\perp jk}} \cos\left\{\varphi_{\|} - \varphi_{\perp} + \phi\right\} \tag{7.14}$$

where all the definitions have been introduced above. The spectral dependence of ϕ was estimated from the magnetic SHG interferometry measurements and was shown to be nearly constant in the wavelength range that corresponds to $\rho_{2\omega}$ maximum. Meanwhile, the relative phase shift $\varphi_{\|} - \varphi_{\perp}$ reveals a resonant-like spectral dependence that was qualitatively similar to $\rho_{2\omega}(\lambda_{2\omega})$, which allows one to associate the resonances in nonlinear magneto-optical response of Co_xAg_{1-x} films with the assistance from LSP modes exited in magnetic nanogranules.

7.6 MSHG in a Disordered Array of Magnetic Nanoparticles

Another example on the magnetization-induced effects in the SHG is presented in this section for the case of an incoherent SHG in disordered arrays of yttrium iron garnet (YIG) nanoparticles in a polymer matrix [31, 32]. YIG ($Y_3Fe_5O_{12}$) as well as YIG-based garnets are well known and intensively studied magneto-optical materials that meet numerous applications, first of all in magneto-optical storage devices. It is worth noting that the first observation on the magnetization-induced SHG effects was performed for YIG films [2].

Apart from the fundamental interest to the problem, nonlinear optics due to its intrinsically high sensitivity can provide effective readout methods for the possible applications in high-density YIG-based magneto-optical memories. At the same time, traditional magneto-optical and magnetization diagnostics methods can be hardly applied for the studies of thin films with small content of magneto-optical material.

In this section we describe some of the experiments on the magnetization-induced SHG in layer-by-layer self-assembled films of YIG nanoparticles (YIG-SA films). Layer-by-layer assembled films containing YIG nanoparticles were deposited on glass substrates by the procedure first suggested in Ref. [31]. They were formed by YIG nanoparticles of the average diameter of about 32 nm coated by a polyelectrolyte. Structural measurements revealed that the films contained separate nanoparticles as well as their aggregates with the average size less than 100 nm.

Figure 7.8a shows the dependence of the absorption on the number of self-assembled layers N measured in the vicinity of the absorption resonance of YIG particles at the wavelength of 350 nm. It can be seen that the absorption is a linear function of N, which proves that the YIG content in each assembled layer is constant. Thus by changing the number of layers we will refer to the controllable change of the number of YIG nanoparticles.

The SHG measurements were performed when using the linearly polarized femtosecond radiation of the Ti–sapphire laser operating at the wavelength of 760 nm; the pump beam was focused onto the sample to a spot of approximately 100 μm in diameter. The angular aperture of the detection system was quite large to gather both the nonlinear optical signal reflected in the direction of the specular reflection and that corresponds to the coherent SHG, as well as the diffuse one that is an evidence of incoherent SHG, or hyper-Rayleigh scattering (HRS) [33]. It was checked that the SHG signal for a polyelectrolyte film without YIG nanoparticles is zero within

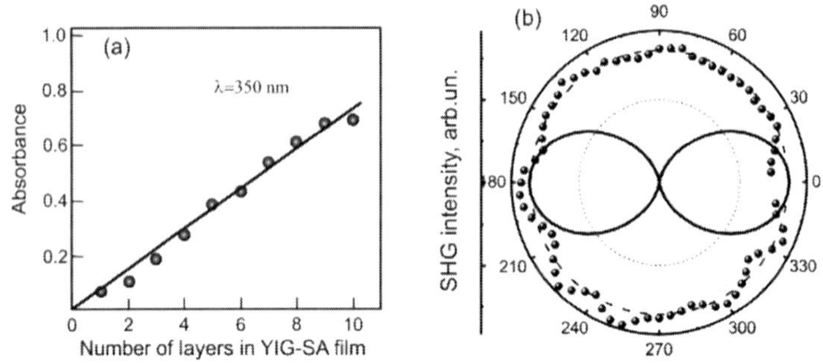

Figure 7.8 (a) Linear optical absorption as a function of the number of layers in YIG-SA films measured for the probe wavelength of 350 nm. (b) SHG intensity polarization diagram for YIG-SA film consisting of seven layers. Solid line: a sample of the polarization diagram for the coherent SHG response; 0 corresponds to the *p*-polarization of the pump and SHG radiation.

the experimental accuracy, and thus the second harmonic from YIG-SA films is generated by YIG particles. Besides, no anisotropy was observed for the SHG response, which is consistent with the in-plane isotropy of the YIG-SA films' structure.

The polarization diagram (symbols in Fig. 7.8b) shows typical dependence of the SHG intensity from YIG-SA films on the azimuthal angle of the analyzer. It is clear from Fig. 7.8b that the SH radiation is depolarized, which is an attribute of HRS. For the comparison, the solid line in the figure shows the polarization dependence that would be expected for a linearly polarized SHG. Depolarization of the SHG from YIG-SA films does not allow us to study the magnetization-induced rotation of the SH polarization plane, so the transversal geometry of the NOMOKE was chosen for the experiment.

First the dependence of the SHG intensity on the number of layers in YIG-SA films was measured in the absence of an external magnetic field, shown in Fig. 7.9 by filled circles. It can be seen that in our particular case of disordered nanoparticles the SHG intensity is a linear function of N, that is, of the number of (nonlinear) particles. This is also typical for the HRS and indicates that the SHG from individual nonlinear dipoles is being summed up incoherently.

For the identification of the NOMOKE in HRS from YIG-SA films the SHG intensity was measured for the opposite directions of the

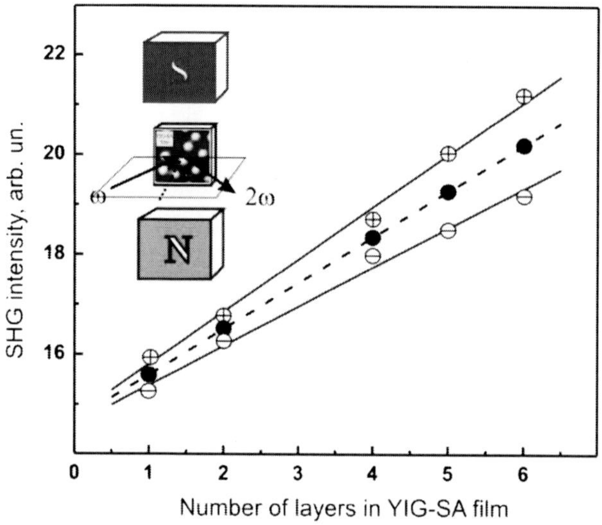

Figure 7.9 Dependences of the SHG intensity measured for a *p*-in, *p*-out combination of polarizations of the SHG and fundamental waves on the number of layers in YIG-SA films measured for the magnetic field $\mathbf{H} = 0$ (black symbols) and for $\mathbf{H} = +2.5$ kOe (⊕) and $\mathbf{H} = -2.5$ kOe (Ø). Solid lines are the approximation by linear dependencies.

magnetic field of 2.5 kOe applied by permanent magnets in the transversal geometry *p*-in, *p*-out combination of polarization was chosen. The corresponding results are shown in Fig. 7.9 (upper and lower curves).

The dependences of the SHG intensity on the number of layers in YIG-SA films measured for the zero, positive, and negative magnetic fields can be approximated by the linear functions of N: $I_{2\omega,i}(N) = K_{2\omega,i}(N) + I_{2\omega,\mathrm{Surf}}$, where the subscript I denotes the used magnetic field strength, $i = 0; +; -$, and the constant $I_{2\omega,\mathrm{Surf}}$ is interpreted as the SHG contribution from the film–air and film–substrate interfaces. Linearity in N along with the observed diffuseness and depolarization is an attribute of HRS, contrary to the coherent SHG where the SHG intensity is quadratic in the thickness of the nonlinear medium (or on the number of the nonlinear dipoles). It can be also seen that application of the magnetic field results in odd with respect to **M** changes in the HRS intensity, while the dependencies $I_{2\omega,\pm}(N)$ remain linear.

In the simplest approximation as we consider linear in **M** expansion of the hyperpolarizability of a single YIG nanoparticle, γ, and neglect the possible tensor character of γ, it is possible to estimate the relative value of the second-order *odd* and *even* components of the hyperpolarizability. Really,

$$\rho_{2\omega} = \frac{I_{2\omega,+} - I_{2\omega,-}}{I_{2\omega,+} + I_{2\omega,-}} = \frac{K_{2\omega,+} - K_{2\omega,-}}{K_{2\omega,+} + K_{2\omega,-}} \approx 2\frac{\gamma^{(2)\text{odd}}_{H=2.5\text{kOe}}}{\gamma^{(2)\text{even}}} \quad (7.15)$$

The experimental data obtained for the films of various thickness give the value $\frac{\gamma^{(2)\text{odd}}_{H=2.5\text{kOe}}}{\gamma^{(2)\text{even}}} \approx 0.07$.

It is worth emphasizing that using the macroscopic technique of the SHG we have obtained averaged information on the relative value of the magnetic susceptibility of YIG nanoparticles or their small aggregates.

Summarizing up, the NOMOKE in second-order HRS from self-assembling films of YIG nanoparticles was studied. In spite of an incoherent character of SHG in such systems, a noticeable magnetic contrast of the NOMOKE was observed.

7.7 Conclusions

Summing up we have shown that magnetization brings about quite noticeable modifications of the nonlinear optical response of magnetic media. Very important is that due to its extremely high surface sensitivity, the NOMOKE and the MSHG can be efficiently applied for the studies of the magnetic properties and of nanostructures. The values of the nonlinear magneto-optical effects are rather huge as compared to linear magneto-optical analogues, that is, driven by the appearance of the nonlinear polarization acting as a source term in the right-hand side of the nonlinear wave equation. We have demonstrated that the symmetry of the magnetization-induced susceptibility components is different from that of nonmagnetic ones and can bring about a strong anisotropy effects in the nonlinear optical response of a structurally isotropic surface. In the case of magnetic granular films of the composition

Co-Ag the role of LSPs excited in Co nanograins in the appearance of the spectral maximum of the NOMOKE is discussed. Finally, magnetization-induced effects in HRS from random arrays of YIG nanoparticles in a polyelectrolyte matrix are described.

Acknowledgments

The authors are thankful to all the members of the research laboratory at the Physics Department of Lomonosov Moscow State university involved in the research on nonlinear magneto-optics, A. A. Nikulin, A. V. Melnikov, T. V. Misuryaev, E. M. Kim, I. A Kolmychek, and I. E. Razdolski, as well as to the colleagues who supplied us with the experimental samples, A. F. Kravets, A. I. Stognij, N. A. Kotov, and coworkers. This work was partially supported by the Russian Foundation for Basic Research.

References

1. O. A. Aktsipetrov, O. V. Braginskii, and D. A. Esikov Magnetization-induced second-harmonic generation in thin garnet films, *Proc ICONO-13* (USSR, Minsk, 1988), 142 (1988).
2. O. A. Aktsipetrov, O. V. Braginskii, and D. A. Esikov Nonlinear optics of gyrotropic media: second harmonic generation in rare-earth iron garnets, *Sov. J. Quantum Electron.*, **20**, 259–263 (1990).
3. N. N. Akhmediev, S. B. Borisov A. K. Zvezdin, I. L. Lyubchanskii, and Yu. V. Melikhov Nonlinear optical susceptibility of magnetically ordered crystals, *Sov. Phys. Solid State*, **27**, 650–652 (1985).
4. R.-P. Pan, H. D. Wie, and Y. R. Shen Optical second-harmonic generation from magnetized surface, *Phys. Rev. B*, **39**, 1229–1234 (1989).
5. J. Reif, J. C. Zink, C.-M. Schneider, and J. Kirschner Effects of surface magnetism on optical second harmonic generation, *Phys. Rev. Lett.*, **67**, 2878–2881 (1991).
6. P. van Gelderen, S. Crampin, Th. Rasing, and J. E. Inglesfield Effect of interface magnetic moments and quantum-well states on magnetization-induced second-harmonic generation, *Phys. Rev. B*, **54**, R2343–R2347 (1996).

7. A. Kirilyuk, Th. Rasing, R. Mégy, and P. Beauvillain Nonlinear magneto-optical response from quantum well states in noble metals: double period and interface localization, *Phys. Rev. Lett.*, **77**, 4608–4611 (1996).
8. H. A. Wierenga, W. de Jong, M. W. J. Prins, Th. Rasing, R. Vollmer, A. Kirilyuk, H. Schwabe, and J. Kirschner Interface magnetism and possible quantum well oscillations in ultrathin Co/Cu films observed by magnetization induced second harmonic generation, *Phys. Rev. Lett.*, **74**, 1462 (1995).
9. O. A. Aktsipetrov, E. M. Kim, R. V. Kapra, T. V. Murzina, A. F. Kravets, M. Inoue, S. V. Kuznetsova, M. V. Ivanchenko, and V. G. Lifshits, Magnetization-induced optical third-harmonic generation in Co and Fe nanostructures, *Phys. Rev. B*, **73**, 140404(R) (2006).
10. T. V. Murzina, T. V. Misuryaev, A. F. Kravets, J. Güdde, D. Schuhmacher, G. Marowsky, A. A. Nikulin, and O. A. Aktsipetrov Nonlinear magneto-optical Kerr effect and plasmon-assisted second-harmonic generation in magnetic nanomaterials exhibiting giant magnetoresistance, *Surf. Sci.*, **482-485**, 1101–1104 (2001).
11. T. V. Murzina, E. M. Kim, S. E. Matskevich, O. A. Aktsipetrov, A. F. Kravets, and A. Y. Vovk Magnetization-induced third harmonic generation in magnetic nanogranular films: correlation with giant magnetoresistance, *JETP Lett.*, **79**, 151–155 (2004).
12. T. V. Murzina, I. A. Kolmychek, A. A. Nikulin, E. A. Gan'shina, and O. A. Aktsipetrov Plasmonic and magnetic effects accompanying optical second-harmonic generation in Au/Co/Au nanodisks, *JETP Lett.* **90**, 504–508 (2009).
13. T. V. Murzina, R. V. Kapra, A. A. Rassudov, O. A. Aktsipetrov, K. Nishimura, H. Uchida, and M. Inoue Magnetization-induced third-harmonic generation in magnetophotonic microcavities, *JETP Lett.*, **77**, 537–541 (2003).
14. O. A. Aktsipetrov, Surface nonlinear optics and nonlinear magneto-optics at Moscow State University, *JOSA B*, **28**, A27–A38 (2011).
15. M. Inoue, K. Arai, T. Fujii, and M. Abe One-dimensional magneto-photonic crystals, *J. Appl. Phys.*, **85**, 5768–5770 (1999).
16. A. A. Fedyanin, T. Yoshida, K. Nishimura, G. Marowsky, M. Inoue, and O. A. Aktsipetrov Magnetization-induced second harmonic generation in magnetophotonic microcavities based on ferrite garnets, *JETP Lett.*, **76**, 527–531 (2002).
17. T. V. Murzina, R. V. Kapra, T. V. Dolgova, A. A. Fedyanin, O. A. Aktsipetrov, K. Nishimura, H. Uchida, and M. Inoue Magnetization- induced second-

harmonic generation in magnetophotonic crystals, *Phys. Rev. B*, **70**, 012407 (2004).

18. Y. R. Shen *The Principles of Nonlinear Optics*, 2nd ed., New York: Wiley, 1984.
19. U. Pustugowa, W. Hübner, and K. H. Bennemann, Enhancement of the magneto-optical Kerr angle in nonlinear optical response, *Phys. Rev. B*, **49**, 10031–10035 (1994).
20. O. A. Aktsipetrov, I. M. Baranova, Yu. A. Ilinskii, Surface contribution to the generation of reflected second harmonic for centrosymmetric semiconductor, *Sov. Phys. JETP*, **64**, 167–173 (1986).
21. A. A. Nikulin and A. V. Petukhov Giant second harmonic on a rough metal surface: fluctuational mechanism of diffusivity and depolarization of radiation, *Sov. Phys. Dokl.*, **34**, 48–52 (s-s 1) (1989).
22. I. A. Kolmychek and T. V. Murzina Magnetizationinduced anisotropy of second harmonic generation in thin cobalt films, *J. Magn. Magn. Mater.*, **323**, 2973–2976 (2009).
23. T. V. Murzina, A. V. Shebarshin, I. A. Kolmychek, E. A. Ganshina, O. A. Aktsipetrov, A. I. Stognij, N. N. Novitski, and A. Stashkevich Magnetism of a cobalt–gold planar nanostructure on the silicon surface, *JETP*, **109**(1), 107–116 (2009).
24. F. Hansteen, O. Hunderi, T. H. Johansen, A. Kirilyuk, Th. Rasing Selective surface/interface characterization of thin garnet films by magnetization-induced second-harmonic generation, *Phys. Rev. B*, **70**, 094408 (7 pages) (2004).
25. A. Kirilyuk Nonlinear optics in application to magnetic surfaces and thin films, *J. Phys. D: Appl. Phys.*, **35**, R189–R207 (2002).
26. F. Hansteen, O. Hunderi, T. H. Johansen, A. I. Kirilyuk, T. H. M. Rasing, Purely magnetization-induced second-harmonic generation in thin garnet films, *Trans Magn. Soc. Jpn.*, **4**, 318–321 (2004).
27. J. Shan, J. I. Dadap, I. Stiopkin, G. A. Reider, and T. F. Heinz Experimental study of optical second-harmonic scattering from spherical nanoparticles, *Phys. Rev. A.*, **73**, 023819 (2006).
28. A. Wokaun, J. G. Bergman, J. P. Heritage, A. M. Glass, P. F. Liao, and D. H. Olson Surface second-harmonic generation from metal island films and microlithographic structures, *Phys. Rev. B.*, **24**, 849 (1981).
29. V. I. Emelyanov and N. I. Koroteev Giant Raman scattering of light by molecules absorbed on the surface of a metal, *Sov. Fisiks-Uspekhi*, **135**, 345 (1981).

30. Yu. G. Pogorelov, G. N. Kakazei, J. B. Sousa, A. F. Kravets, N. A. Lesnik, M. M. Pereira de Azevedo, M. Malinowska, and P. Panissod Structural and magnetic study of heterogeneous Co_xAg_{1-x} films by resonance and magnetometric techniques, *Phys. Rev. B*, **60**, 12200–12206 (1999).
31. J. W. Ostrander, A. A. Mamedov, and N. A. Kotov Two modes of linear layer-by-layer growth of nanoparticle-polylectrolyte multilayers and different interactions in the layer-by-layer deposition, *J. Am. Chem. Soc.*, **123** 1101–1110 (2001).
32. O. A. Aktsipetrov Nonlinear magneto-optics in magnetic nanoparticles, *Colloids Surf. A*, **202**, 165–173 (2002).
33. A. V. Melnikov, A. A. Nikulin O. A. Aktsipetrov, HyperRayleigh scattering by inhomogeneous thin films of Pb*x*Zr0.53Ti0.47O3: Disorder effects, *Phys. Rev. B.*, **67**, 134104 (9 pages) (2003).
34. O. A. Aktsipetrov, T. V. Murzina, E. M. Kim, R. V. Kapra, A. A. Fedyanin, M. Inoue, A. F. Kravets, S. V. Kuznetsova, M. V. Ivanchenko, and V. G. Lifshits Magnetization-induced second and third-harmonic generation in magnetic thin films and nanoparticles, *J. Opt. Soc. Am. B*, **22**, 137–& (2005).
35. K. H. Bennemann Theory for nonlinear magneto-optics in metals, *J. Magn. Magn. Mater.*, **200**, 679–705 (1999).
36. R. R. Birss *Symmetry and Magnetism, Series of Monographs on Selected Topics in Solid State Physics*, 2nd ed. Amsterdam: North-Holland, 1966.
37. B. Koopmans, M. G. Koerkamp, Th. Rasing, and H. van den Berg Observation of large Kerr angles in the nonlinear optical response from magnetic multilayers, *Phys. Rev. Lett.*, **74**, 3692–3695 (1995).
38. T. V. Murzina, A. A. Nikulin, O. A. Aktsipetrov, J. W. Ostrander, A. A. Mamedov, N. A. Kotov, M. A. C. Devillers, and J. Roark Nonlinear magneto-optical Kerr effect in hyper-Rayleigh scattering from layer-by-layer assembled films of yttrium iron garnet nanoparticles *Appl. Phys. Lett.*, **79**, 1309–1311 (2001).
39. Th. Rasing Nonlinear magneto-optical studies of ultrathin films and multilayers, in *Nonlinear Optics in Metals*, 132–218, K. H. Bennemann, Ed. Oxford, UK: Clarendon, 1998.

Chapter 8

Magnetoplasmonic Nanostructures

César Clavero and Rosa Alejandra Lukaszew
Department of Applied Science and Department of Physics, College of William and Mary, P.O. Box 8795, Williamsburg, VA 23187, USA
cclavero@wm.edu

The possibility of concentrating and manipulating light at unprecedentedly low size scales by means of surface electromagnetic waves such as surface plasmon polaritons (SPPs) or localized surface plasmons (LSPs), well below the diffraction limit, opens a strikingly broad spectrum of fundamental and practical applications such as plasmonic circuits, plasmonic lenses, light trapping, biosensors, and cancer treatment, among many others. Nevertheless, for many of these applications it is important to have external control of the propagation of such plasmonic resonances, for example, using external magnetic fields. In this chapter we will discuss how adequately inserting magneto-optically active metals such as ferromagnetic transition metals in systems with plasmonic properties such as thin films, multilayers, and nanostructures imparts to them a strong sensitivity to external magnetic fields. In turn these systems exhibit also a remarkable increase of the magneto-optical activity due to interaction of the strong electromagnetic fields generated by the plasmonic excitations and the magneto-optically active materials.

Handbook of Nanomagnetism: Applications and Tools
Edited by Rosa A. Lukaszew

ISBN 978-981-4613-04-0 (Hardcover), 978-981-4613-05-7 (eBook)
www.panstanford.com

8.1 Introduction

Materials with free electrons such as metals, conducting oxides, and semiconductors can sustain plasmonic excitations due to collective oscillations of the conduction electrons in the material in response to external electromagnetic fields. In particular, the so-called surface plasmon polariton (SPP) modes can appear in the case of flat or corrugated interfaces illuminated under suitable conditions. SPP modes are transverse magnetic (TM) surface waves propagating at the interface between materials with opposite signs of the real part of their dielectric permittivities, as it is, for example, the case of a metal and a dielectric [1]. Since they result from the interactions between light and the free electrons of the conductor on the surface, they generate very intense and highly confined electromagnetic fields at the interface. These plasmonic modes are frequently referred to as a way to squeeze and concentrate light well below the diffraction limit due to the extreme field concentration that takes place [2]. The wave vector $\mathbf{k}_{\mathrm{sp}}$ of a SPP mode propagating at the interface between a metal and a dielectric can be expressed as [1]

$$\mathbf{k}_{\mathrm{sp}} = \mathbf{k}_0 \sqrt{\frac{\varepsilon_{\mathrm{m}}\varepsilon_{\mathrm{a}}}{\varepsilon_{\mathrm{m}} + \varepsilon_{\mathrm{a}}}} \tag{8.1}$$

where $\mathbf{k}_0 = \omega/c$ represents the wave vector of light in vacuum and ε_{m} and ε_{a} are the dielectric constants of the metal and the dielectric, respectively. The dispersion relation for SPP modes propagating on Au, a good plasmonic material due to its low optical absorption in the visible range, is shown in Fig. 8.1a, considering it a perfect Drude metal ($\omega_{\mathrm{p}} = 9$ eV, $\Gamma = 0.067$ eV). Due to their bound nature, SPP modes appear to the right of the light line in the dielectric medim and up to the characteristic surface plasmon frequency $\omega_{\mathrm{sp}} = \omega_{\mathrm{p}}/\sqrt{1+\varepsilon_{\mathrm{a}}}$, where ω_{p} is the plasma frequency of the metal. The wave vector of SPP modes $\mathbf{k}_{\mathrm{sp}}$ approaches a finite limit at the surface plasmon frequency ω_{sp} due to free electron and interband damping in the material, limiting the propagation length of SPP modes to $L = (2Im[\mathbf{k}_{\mathrm{sp}}])^{-1}$, typically between 10 and 100 μm in the visible range, depending on the metal/dielectric combination [3]. Radiation into the metal occurs in the transparency regime above ω_{p}. Also, due

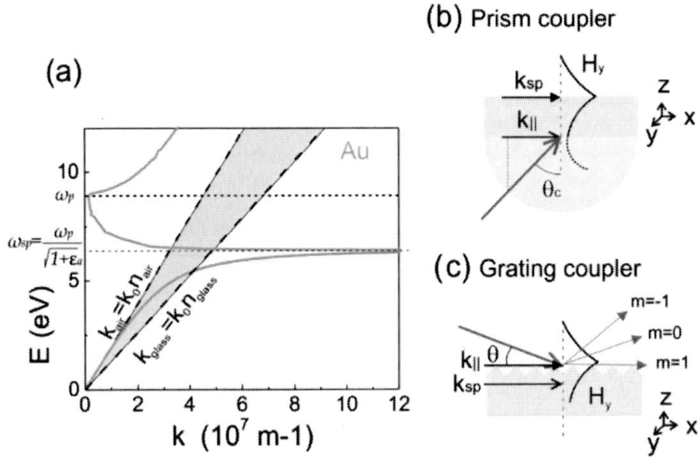

Figure 8.1 (a) Dispersion relation for SPP modes propagating on the surface of a Au film, considering it a perfect Drude metal ($\omega_p = 9$ eV, $\Gamma = 0.067$ eV) in contact with air. (b) Prism coupler configuration (Kretschmann configuration) and (c) grating coupler configuration. k_{sp} represents the momentum of SPP modes, $k_{||}$ the parallel component of the momentum for the incident light, and H_y the SPP-generated magnetic field.

to damping in the metal, the *quasibond* leaky part of the dispersion relation among ω_{sp} and ω_p is allowed [3].

From Eq. 8.1 it can be inferred that the momentum for the SPP modes is always greater than that of a photon propagating through the dielectric medium. It is thus necessary to enhance the momentum of the incoming light to match the momentum of the SPP modes using some kind of coupling mechanism. The most commonly used configurations to achieve optical excitation of SPP modes are (i) prism coupler–based structures (also known as the Kretschmann configuration) (Fig. 8.1b) where the metallic films are optically coupled to a prism and far-field light is angled through the prism such that the in-plane wave vector $\mathbf{k}_{||}$ matches the associated SPP propagation constant $\mathbf{k}_{sp}$, and (ii) diffraction by gratings (Fig. 8.1c) in which the diffracted beams with increased momentum couple with SPP modes on the surface. The coupling of light to SPP modes can be easily observed in both geometries as a dip in the angular or wavelength spectrum of the reflected light due to SPP absorption.

SPP modes have been extensively studied and used in numerous applications, including light concentration below the diffraction limit [4], plasmonic circuits [5, 6], light trapping in photovoltaic cells [7], transmittance modulation in smart windows [8], biosensors [9], and cancer treatment [10] among many others. Nevertheless, for many of these applications it is desirable to have some kind of external control of the SPP modes propagation by means of, for example, external magnetic fields. Early theoretical works back in the early 1970s [11–13] described the effect of externally applied magnetic fields on the propagation of surface plasmons on metals or semiconductors, introducing for the first time the concept of magnetoplasmic surface waves. Ferguson et al. [14] demonstrated experimentally the presence of such modification in the propagation of SPP modes on ferromagnetic Ni thin films using the prism coupler configuration and applying alternate magnetic fields perpendicular to the incident plane, in the so-called transverse configuration. They reported for the first time the modification of the SPP wave vector $\mathbf{k}_{sp}$ due to external magnetic fields, accompanied by a remarkable enhancement of the magneto-optical activity [15], explaining similar previous observations by Judy et al. [16, 17] on Fe thin films. Interactions of SPP modes with external magnetic fields have been predicted and observed in multiple subsequent works in ferromagnetic thin films and ferromagnetic-noble metal multilayers [18–25]. Even though the most commonly used SPP generation technique has been the prism coupler configuration, modulation of surface plasmons generated by means of grating coupling has also been demonstrated [26, 27]. The propagation of SPP modes, and thus their dispersion relation, is only affected by the externally applied magnetic fields for the case of the transverse configuration [28–30], that is, a magnetic field applied parallel to the interface and perpendicular to the SPP wave vector $\mathbf{k}_{sp}$ (Fig. 8.2a). For this particular case, a splitting of the dispersion relation for SPP modes is observed when magnetic fields are applied in opposite directions. Figure 8.2b shows the dispersion relation of SPP modes propagating on cobalt and gold thin films in contact with air when a magnetic field is applied along the transverse configuration in opposite directions, calculated by using a matrix transfer formalism [31] and imposing the boundary conditions for SPP modes

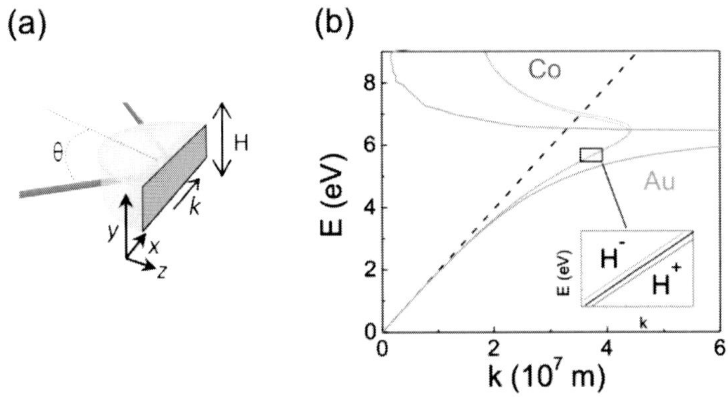

Figure 8.2 (a) Prism coupler transverse configuration. (b) Dispersion relation for SPP modes in Co and Au films in contact with air, considering their absorption losses. Due to its magneto-optical activity, the dispersion relation of Co splits into two branches when external magnetic fields are applied in opposite directions along the *y* direction, in the so-called transverse configuration.

[19, 32]. A clear splitting of the dispersion relation is observed for the case of the ferromagnetic cobalt thin film when applying low magnetic fields around 20 mT, demonstrating external control of the propagation characteristics of SPP modes. Much higher fields up to several teslas would be necessary to obtain an observable change in the dispersion relation of Au [33].

As previously mentioned, modification of SPP modes can only be achieved in the transverse configuration, where the externally applied magnetic field is perpendicular to the SPP wave vector $\mathbf{k}_{sp}$. Nevertheless, a significant enhancement of the magneto-optical activity is observed in all the three possible orientations of the magnetic field with respect to the incidence plane, that is, transverse, longitudinal, and polar, due to the intense SPP-generated electromagnetic fields that penetrate the ferromagnetic material [21–23, 34].

The control and modulation of SPP modes by external magnetic fields has led to striking results, such as the manipulation of optical transmission through magnetoplasmonic crystals [35–39], the control of terahertz SPP modes on semiconductor surfaces by applying an external static magnetic field [40], and a remarkable enhancement of the magneto-optical response [23, 34, 41, 42]. As a consequence, a number of technological applications have appeared in the last few years, including magnetoplasmonic optical switches and modulators [43, 44], optical isolators [45], plasmonic circuits [46], and magnetoplasmonic sensors [47].

In this chapter, special attention will be devoted to SPP mode modulation in noble-ferromagnetic multilayers such as Au/Co/Au using the prism coupler configuration, which allows for intense SPP modes highly sensitive to external magnetic fields [21–23, 34]. Also, less stringent geometries will be explored as it is the case of Au–Co nanocomposite thin films, in which the magneto-optical active medium is spread through the whole thickness of the film in the form of Co nanoparticles embedded in a Au matrix, allowing for further optimization of the modulation of SPP modes [48]. Also other coupling geometries will be discussed as it is the case of noble-ferromagnetic metal multilayers with grating morphology [27], which also allow for excitation and modulation SPP modes in a fundamentally different incoming light–SPP interaction geometry.

Nevertheless, there is a second fundamental plasmonic excitation that can also appear in appropriately designed metallic and metal–dielectric nanostructures known as localized surface plasmons (LSPs). LSP modes are nonpropagating excitations of the conduction electrons of metallic or semiconductor nanostructures coupled to the incoming electromagnetic field [3]. They impart unusual optical properties to these nanostructures, such as strongly enhanced size-, shape-, and medium-dependent light absorption and strong and highly localized electromagnetic fields [49]. Nanostructures supporting LSP modes have been employed in a wide range of applications, including surface-enhanced Raman spectroscopy [50], biosensing [51, 52], biological imaging [53], cancer treatment [2], plasmonic nanoantennas [54], and photovoltaic devices [7] among many others. Nevertheless, these localized modes also exhibit sensitivity to external magnetic fields when magneto-optically active materials such as ferromagnetic metals are incorporated in their structure. In addition, due to the fact that LSP modes generate very intense electromagnetic fields, they give rise to an unprecedented enhancement of the magneto-optical activity in these systems. In previous studies, strong magneto-optical effects

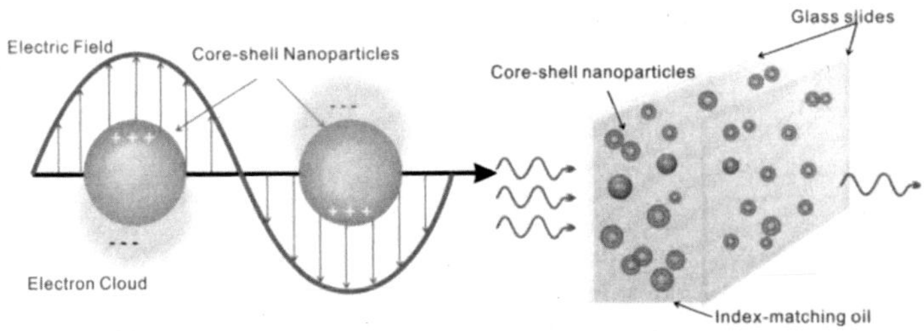

Figure 8.3 (a) Scheme of localized surface plasmons (LSPs) in core–shell nanoparticles. The displacement of the conduction electron cloud relative to the nuclei is schematically shown. (b) Scheme of core–shell Co–Ag nanoparticles suspended in refractive index-matching oil ($n = 1.5018$) between two glass slides. Adapted from Ref. [42].

and control of the optical transmittance has been described in systems with periodically distributed metallic and dielectric structures due to the excitation of LSP modes [38, 55–57]. Also, a strong enhancement of the magneto-optical response has been observed in ferromagnetic nanoparticles embedded in dielectric matrices [58–63], nickel nanowire arrays [57, 64], yttrium–iron–garnet films with Au nanoparticles embedded [65], magnetic thin films with Au nanoparticles on top [66, 67], magnetite nanocrystals near Au surfaces [68], and noble metal–ferromagnetic metal nanosandwich structures [33, 69–73] at those frequencies where LSP modes are excited. Of especial interest are core–shell nanoparticles formed by ferromagnetic metals in the core surrounded by noble metals (Fig. 8.3). The strong electromagnetic fields generated by LSP excitation in the shell penetrate the ferromagnetic core and give rise to a remarkable enhancement of the magneto-optical activity [41, 42, 74] due to LSP excitation. In this chapter we will discuss more in depth the properties of core–shell nanoparticles. Interestingly, interaction between the two plasmonic modes previously described, LSP and SPP, has been described, as, for example, in Au/Co/Au trilayers with gold nanodiscs patterned on top [75, 76] and in ferromagnetic nanoparticles capped with Au–dielectric layers [77].

Interaction of plasmonic excitations with magnetism opens a new venue for controlling and manipulating light at unprecedentedly small scales, yielding a very broad spectrum of future applications. In this chapter we will discuss some of the most representative systems that allow control of SPP and LSP modes by means of external magnetic fields and that in most of the cases exhibit a considerably enhanced magneto-optical response due to the generated enhanced electromagnetic fields.

8.2 Thin Films and Multilayers

8.2.1 *Multilayers*

As previously mentioned, modification of SPP modes propagating on materials with free electrons is possible by applying external magnetic fields perpendicular to the SPP wave vector $\mathbf{k}_{sp}$. Nevertheless, for the case of noble metals such as gold and silver in which SPP excitation is optimum in the visible range, magnetic fields of the order of several teslas are necessary to give rise to noticeable changes in $\mathbf{k}_{sp}$, making its application impractical [24]. On the other hand, ferromagnetic metals such as iron, cobalt, or nickel have a much larger magneto-optical activity and allow modification of SPP modes at low magnetic fields [14, 15]. Nevertheless, they exhibit an elevated optical absorption giving rise to overdamped short-range SPP modes with low associated electromagnetic fields. To overcome these drawbacks, multilayers composed by noble and ferromagnetic metals have been proposed as a prototypical system that combines intense SPP-generated electromagnetic fields, high dependence of the SPP $\mathbf{k}_{sp}$ vector with external magnetic fields, and high magneto-optical activity [21–24, 34, 78]. Different combinations of materials and thicknesses have been investigated in the literature [34, 47, 79–83]; nevertheless Au–Co–Au has been the most frequently used [21–23, 84]. Figure 8.4a shows the angular dependency of the reflectivity in the prism coupler configuration for Au and Co thin films with critical thicknesses at which optimum coupling is achieved, that is, a zero in reflectivity is obtained at a certain angle [1]. A much sharper minimum is found for Au due to its low optical absorption as compared to Co, which is also associated with much more intense SPP-generated electromagnetic fields on the upper metal–dielectric interface [1], as shown in Fig. 8.4b.

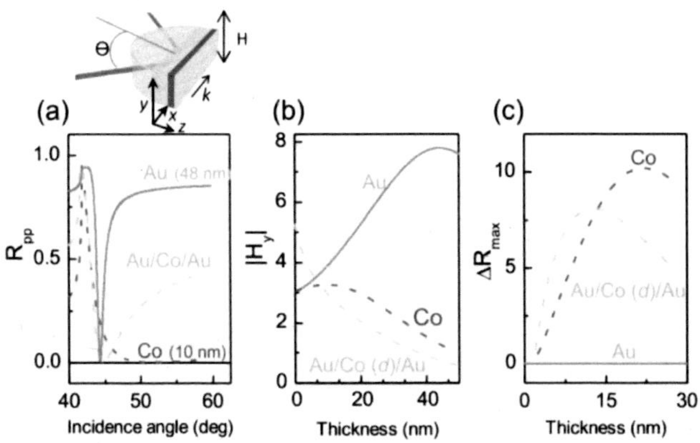

Figure 8.4 (a) Angular dependency of the reflectivity for a 48 nm Au film, a 10 nm Co film, and a Au (3 nm)/Co (3 nm)/Au (20 nm) multilayer. (b) SPP-generated $|H_y|$magnetic field at the top metal–air interfaces. (c) Transverse magneto-optical Kerr effect ΔR as a function of thickness for a Co film and for Au (3 nm)/Co (d_{Co})/Au (20 nm) trilayers.

The design of Au–Co–Au multilayers with adequate properties is challenging. The ferromagnetic film needs to be placed as close as possible to the upper interface to take advantage of the high SPP-generated electromagnetic fields [23]. Nevertheless, due to the tendency of ferromagnetic metals to get oxidized in atmospheric conditions, a thin Au capping layer is deposited on top to preserve its properties. Also, the total thickness of the trilayer needs to be tuned to achieve critical thickness and thus optimum SPP excitation in the prism coupler configuration. Figure 8.4a shows the reflectivity curve for a selected Au (3 nm)/Co (3 nm)/Au (20 nm) multilayer satisfying all these conditions. While Au–Co–Au trilayers exhibit broader reflectivity curves as compared to Au films due to the presence of Co (Fig. 8.4a), the electromagnetic fields in the top metal–air interfaces are considerably enhanced for Co thicknesses below 7.5 nm (Fig. 8.4b). Strikingly, due to the strong and highly confined SPP-generated electromagnetic fields that penetrate the ferromagnetic Co film in the Au/Co/Au trilayers, the magneto-optical response in the transverse configuration $\Delta R = R(^{+}\mathbf{H}) - R(^{-}\mathbf{H})$ is strongly enhanced for Co thicknesses below 13.5 nm as compared to single Co films with the same thicknesses [21, 34] (Fig. 8.4c).

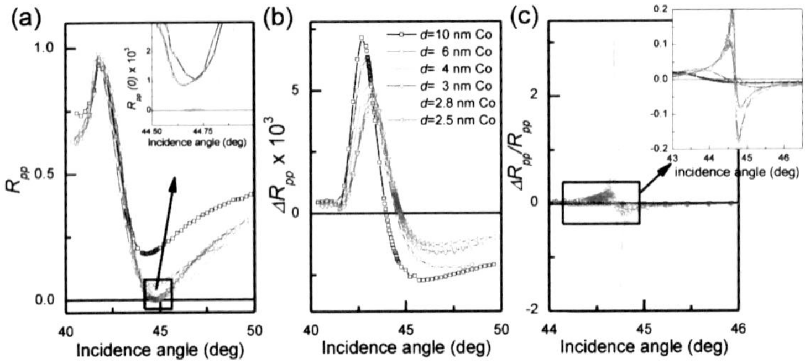

Figure 8.5 (a) Angular dependency of the reflectivity with no external magnetic field applied R_{pp}, (b) transverse magneto-optical Kerr effect $\Delta R = R(+\mathbf{H}) - R(-\mathbf{H})$ measured in the Kretschmann configuration, and (c) relative variation in reflectivity $\Delta R_{\text{pp}}/R_{\text{pp}}$ for the Au (3 nm)/Co (d)/Au (20 nm) trilayers, with d ranging from 2.5 to 10 nm. Adapted from Ref. [23].

As deduced from Fig. 8.4, the ferromagnetic film thickness within the multilayers is the most critical parameter to obtain the desired properties and makes challenging the experimental fabrication of these systems. Figure 8.5a shows reflectivity curves for Au (3 nm)/Co (d_{Co})/Au (20 nm)/glass multilayers, with Co thickness d_{Co} ranging from 2.5 to 10 nm fabricated by means of sputtering deposition on glass [23, 85]. Clearly, a thickness control down to the atomic level is needed to obtain the critical thickness and thus a zero in the reflectivity as shown in the inset to Fig. 8.5a. For this particular system, this condition is satisfied for Co thicknesses around 2.8 nm Co. A continuous increase of the transverse magneto-optical response ΔR is obtained as the thickness of the ferromagnetic films is increased, as shown in Fig. 8.5b. Nevertheless, it is worth noticing that the transverse magneto-optical activity is normally expressed as relative variation of the reflectivity $\Delta R/R$. Since a zero in the reflectivity can be obtained for carefully tailored trilayer thicknesses, this gives rise to an artificial enhancement of the $\Delta R/R$ ratio (Fig. 8.5c), yielding a very sharp and sensitive signal [23, 47]. In particular, the Au–Co–Au trilayer with $d_{\text{Co}} = 2.8$ nm shown in Fig. 8.5b exhibits a maximum $\Delta R/R$ of 320%, a value considerably larger than that found in Co thin films with no SPP excitation around 0.5% [22]. This very intense and sharp signal has outstanding applications for example in biosensing [47].

Another interesting characteristic that systems with magneto-plasmonic properties offer is the possibility of actually probing the electromagnetic field distribution inside the metal trilayer with a very high spatial resolution. The SPP-generated electromagnetic fields can be probed by positioning a ferromagnetic thin film in different positions within the multilayer and measuring the magneto-optical response [23] since it is directly proportional to the intensity of the electromagnetic field in the ferromagnetic material [21, 34, 86]. Experimental measurements of this kind have been done in multilayer systems [21, 23, 24, 33], Au/Co/Au nanodiscs [71, 73], and Au/Co/Au trilayers with gold nanodiscs patterned on top [75, 76].

For the case of SPP modes propagating on a thin film or multilayer, the variation of the complex reflection coefficient for p-polarized light Δr_{pp} in the transverse configuration is directly proportional to the product of the TM electric field components ($\mathbf{E}_x$ and $\mathbf{E}_z$) within the magnetic layer [78], which along with $\mathbf{H}_y$ are the only electromagnetic fields present in TM modes. In the ultrathin film limit, where the electric fields can be considered constant within the magnetic film, this relationship is given by [21, 86]

$$\Delta r_{\text{pp}} = \frac{i\mathbf{k}_0^2}{2\kappa_1 t}\varepsilon_{xy} l_{\text{m}} \mathbf{E}_x \mathbf{E}_z \tag{8.2}$$

where $\mathbf{k}_0 = \omega/c$ is the wave number in vacuum, κ_1 is the z component of the incident light wave vector at the first interface, ε_{xy} are the magneto-optical constants of the magnetic film, l_{m} is the thickness of the magnetic film, and t accounts for the absorption of such film [21]. Thus, by placing a magneto-optically active layer in different positions within the Au film and measuring the variation of the complex reflectivity, it is possible to probe the actual intensity of the SPP-generated electromagnetic fields at the location of the magnetic layer. This study has been carried out for example for Au/Co/Au trilayers with a 2.8 nm thick Co layer positioned at different distances from the upper air–Au interface [23], as shown in Fig. 8.6. Reflectivity and magneto-optical measurements in different configurations were carried out to extract the change in the complex reflection coefficient Δr_{pp} when an external magnetic field is applied

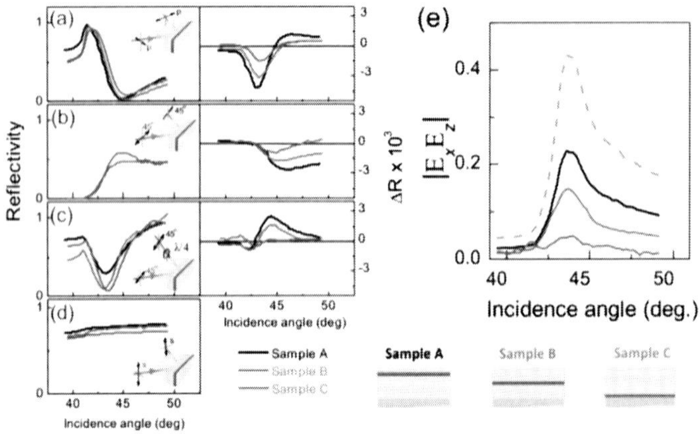

Figure 8.6 Reflectivity and $\Delta R = R(+\mathrm{H}) - R(-\mathrm{H})$ magneto-optical response angular dependence for Au/Co/Au trilayers with Co positioned at 3 nm (A), 11.5 nm (B), and 23 nm (C) to the upper air–Au interface using (a) *p*-polarized light, (b) light rotated 45°, (c) light rotated 45° and a quarter wave plate, and (d) *s*-polarized light. (e) Modulus of the product of the TM electric components $|\mathrm{E}_x \mathrm{E}_z|$ normalized to the incident intensity at the position in which the ferromagnetic film is placed in each case for the trilayers. Adapted from Ref. [23].

[21]. For this purpose *p*-polarized light was used (Fig. 8.6a), light rotated 45° from the *p*axis and with a polarization analyzer oriented at 45° (Fig. 8.6b), light rotated 45° from the *p*axis and with a quarter wave plate with its fast axis along *p* followed by a polarization analyzer oriented at 45° (Fig. 8.6c), and *s*-polarized light (Fig. 8.6d). Strong differences in the magneto-optical response are observed depending on the position of the ferromagnetic film. The variation of the complex reflection coefficient Δr_{pp} and thus the product of the TM electric field components ($\mathbf{E}_x$ and $\mathbf{E}_z$) as a function of the position within the Au films can be obtained, properly combining the experimental data shown in Fig. 8.6a-d. Figure 8.6e shows the measured $\mathbf{E}_x\mathbf{E}_z$ fields when positioning the Co film at different distances to the upper Au–air interface, normalized to the incident radiation. The results clearly show an exponential decay in the SPP modes generated electromagnetic fields as the distance to the upper Au–air interface increases. The calculated intensity of the

electromagnetic field at the upper interface is shown with a dashed line in Fig. 8.6e for reference.

As shown in this section, multilayers formed by noble and ferromagnetic metals are prototypical systems in which the propagation of SPP modes is affected by external magnetic fields and a strong enhancement of the magneto-optical response is observed due to the SPP-generated electromagnetic fields that, as it has been shown, can be probed. In the following sections, it will be discussed other systems with similar properties but with a different structure like Au–Co nanocomposite films and multilayers with a corrugated structure.

8.2.2 *Nanocomposite Thin Films and Multilayers*

From the discussion in the previous section it becomes clear that to obtain an optimum response in noble metal–ferromagnetic metal multilayers it is critical to achieve a very accurate control on the thin-film thicknesses. The possibility of using nanocomposite noble metal–ferromagnetic metal films such as gold–cobalt has been proposed [48] in order to explore less stringent material designs that still allow external control of SPP propagation. Interestingly, gold and cobalt do not form a binary alloy [87] and thus they are ideal candidates to fabricate nanocomposite materials prepared by means of physical vapor deposition methods such as sputtering codeposition. The two metals form different nanocluster–matrix aggregates, depending on their relative concentration and the growth temperature, and thus their magnetoplasmonic properties can be optimized. Figure 8.7a,b shows scanning transmission electron microscopy (STEM) images for two selected samples, namely Co 5% Au 95% deposited at 300°C and Co 40% Au 60% deposited at 600°C. Both images show homogeneous segregation of Au and Co forming clear grains in the perpendicular and lateral directions. Nevertheless, it becomes clear how the deposition temperature allows tuning of the size of the Co nanoparticles, giving rise to larger Co grains for higher deposition temperatures. Multiple factors such as surface morphology, Co concentration, nanoparticle size, and interparticle distance strongly condition the optical and magneto-optical response of nanocomposite films [48].

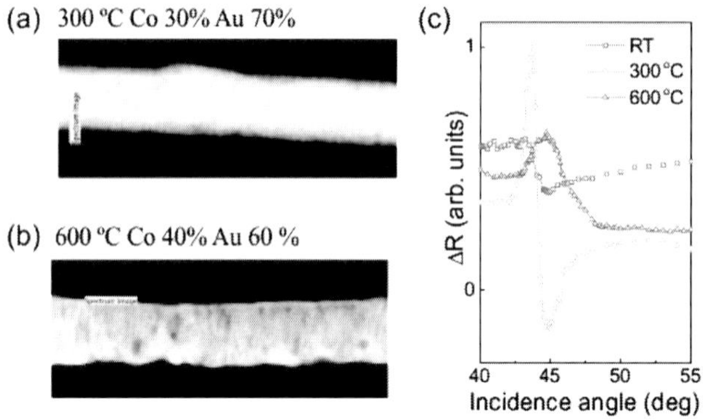

Figure 8.7 STEM images for (a) Co 30% Au 70% nanocomposite film grown at 300°C and (b) Co 40% Au 60% nanocomposite film grown at 600°C. (c) Angular dependence of the transverse magneto-optical Kerr effect $\Delta R = R(\mathrm{H}^+) - R(-\mathrm{H})$ for Co 10% Au 90% nanocomposite films grown at different temperatures. Adapted from Ref. [48].

In particular, ellipsometry measurements demonstrate that only the optical response for the films grown at 300°C can be satisfactorily modeled using the Bruggeman effective medium formalism [88], which is indicative of a good segregation of Co and a homogeneous distribution in the Au matrix for that particular growth temperature independently of the Co concentration. Figure 8.7c shows the transverse magneto-optical Kerr effect $\Delta R = R(\mathbf{H}^+) - R(\mathbf{H}^-)$ measured in the prism coupler configuration for nanocomposite films with a Co 10% Au 90% concentration ratio and grown at room temperature, 300°C, and 600°C. The maximum magneto-optical effect is found for films grown at 300°C, for which optimum segregation of Co in the Au matrix was inferred by the optical analysis. This fact indicates that for this particular morphology the SPP-generated electromagnetic fields are able to penetrate the individual Co nanoparticles and strongly enhance the magneto-optical response. Too small and diluted nanoparticles such as the ones obtained by deposition at room temperature or too large nanoparticles giving rise to an excessively rough surface as the ones grown at 600°C hinder the interaction between SPP modes and Co. Even though the main contribution to the magneto-optical response for the case of the prims coupler configuration is due to

SPP excitation and modulation, contributions due to LSP excitation and interaction between the Co nanoparticles cannot be discarded, as will be shown in Section 8.3.

8.2.3 *Multilayers on Gratings*

As previously shown in Fig. 8.1c, SPP modes can also be generated by using other coupling methods such as periodically modulated interfaces and diffraction gratings [1], allowing for new ways of controlling externally the propagation of SPP modes [27, 56]. This geometry changes fundamentally the way in which light and SPP modes interact since in this case the incoming radiation illuminates directly the surface of the metal. Thus, the strict film thickness constrains previously discussed for the prism coupler configuration are not relevant any more. In addition this method further expands the possibilities of tailoring the SPP excitation and propagation since parameters such as grating material, period, and shape will strongly determine the nature of the excited SPP modes.

Preliminary works by Newman et al. [26] showed that far-field diffused light originated from SPP modes propagating on Ni gratings and recoupling intro the incident medium varies slightly when an external magnetic field is applied in the transverse geometry. Nevertheless the high optical absorption of Ni gives rise to weak SPP modes with low associated electromagnetic fields. Subsequent studies showed how the wave vector $\mathbf{k}_{sp}$ for intense SPPs excited through a periodic modulation of the upper surface in Au/Co bilayers and Au/Co/Au trilayers can be externally modified by applying modest external magnetic fields also in the transverse configuration [27]. Following to these studies, several groups have described considerable enhancements in the magneto-optical response of periodic structures due to SPP propagation [89, 90] and in the so-called magnetoplasmonic crystals [56].

Let us consider a grating which grooves are oriented perpendicular to the light incidence plane (Fig. 8.8a). In addition to the reflected beam, also known as zero diffraction order ($m = 0$), higher diffracted orders are observed ($m = \pm 1, \pm 2, \ldots$) at different angles, depending on parameters such as incident wavelength λ, incidence angle θ, or grating period Λ (Fig. 8.8a). The parallel wave

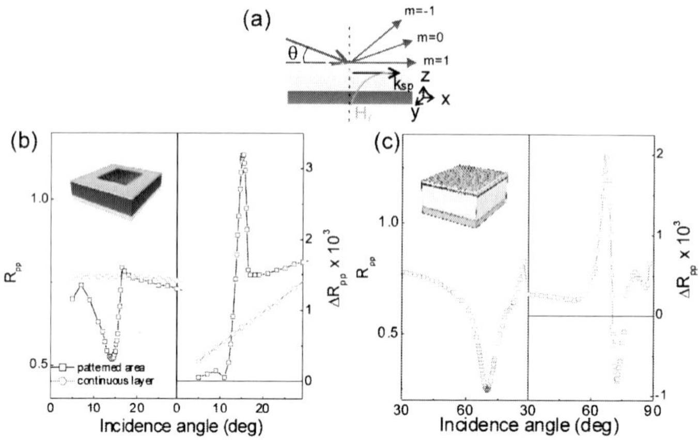

Figure 8.8 (a) Schematic view of momentum matching by a grating. (b) Reflectivity (left) and TMOKE $\Delta R_{pp} = R_{pp}(H^{+}) - R_{pp}(H^{-})$ (right) for a Au (5 nm)/Co (50 nm) bilayer with a PMMA grating lithographed on top. (c) Reflectivity (left) and TMOKE ΔR_{pp} for a Au (3 nm)/Co (3.5 nm)/u (80 nm) trilayer deposited on a polycarbonate grating. Adapted from Ref. [27].

vector component $\mathbf{k}_{||}$ of the incoming light is increased by an amount related to the grating period Λ when one of the higher diffracted orders yields parallel to the surface

$$\mathbf{k}'_{||} = \mathbf{k}_{||} + m\frac{2\pi}{\Lambda} \tag{8.3}$$

Under such circumstances, excitation of SPP modes is possible for those incident angles that satisfy $\mathbf{k}'_{||} = \mathbf{k}_{sp}$, where $\mathbf{k}_{sp}$ is the SPP wave vector given by Eq. 8.1 for the case of a metal–dielectric interface. The wave vector $\mathbf{k}_{sp}$ for SPP modes excited through a periodic modulation can also be modified by applying external magnetic fields in the transverse configuration, as demonstrated in Au/Co bilayers and Au/Co/Au trilayers [27], in analogy to the results presented in Section 8.2.1 for the prims coupler configuration. This effect is accompanied by a strong enhancement of the magneto-optical response. To probe this concept, a poly(methyl methacrylate) (PPMA) grating was lithographed on top of a Au (5 nm)/Co (50 nm) bilayer covering a 1 mm^2 area, thus allowing for excitation of SPP modes on the upper surface, as demonstrated by the dip in the p reflectivity R_{pp} shown in Fig. 8.8b at 13° when illuminating inside the patterned area [27]. In addition, strong variations of the measured reflectivity are detected around that angle when a 30 mT magnetic field is applied in the transverse configuration and in opposite directions $\Delta R_{pp} = R_{pp}(\mathbf{H}^{+}) - R_{pp}(\mathbf{H}^{-})$. Such strong variations are due to the modification of $\mathbf{k}_{sp}$ by the external magnetic field, which in turn modifies the SPP coupling condition and shifts the dip in the reflectivity curve. This result demonstrates unambiguously that SPP modes propagating on a grating can be externally modified. In analogy with the trilayers shown in Section 8.2.1, this concept was also shown to work for the case of Au/Co/Au trilayers deposited on polycarbonate gratings, where the different films grow conformably to the template underneath, thus transmitting it through the different films up to the top surface (Fig. 8.8c). In this case, again a dip in the reflectivity is associated with a considerable enhancement of the magneto-optical response ΔR_{pp} due to modulation of SPP modes. Different design factors such as grating shape and pitch, ferromagnetic film thickness and position within the trilayer can be further optimized in order to maximize the modulation of SPP modes and the magneto-optical response of the system [27].

8.3 Core–Shell Nanoparticles

In previous sections, the effect of applying external magnetic fields on SPP modes propagating on flat or corrugated surfaces has been reviewed. Nevertheless, there is a second fundamental plasmonic excitation known as LSP consisting of nonpropagating excitations of the conduction electrons that appear in appropriately designed metallic and metal–dielectric nanostructures. Similarly to SPP excitation, LSP is accompanied by highly localized and intense electromagnetic fields that can impact the magneto-optical activity of ferromagnetic nanoparticles embedded in dielectric matrices [58–62], nickel nanowire arrays [57, 64], yttrium–iron–garnet films with Au nanoparticles embedded [65], Au–Fe nanoparticles in a SiO_2 matrix [91], ferromagnetic nanodisks [70], noble metal–ferromagnetic metal nanosandwich structures [69–71], magnetic films in proximity to gold nanoparticles [67, 75], nanoparticle

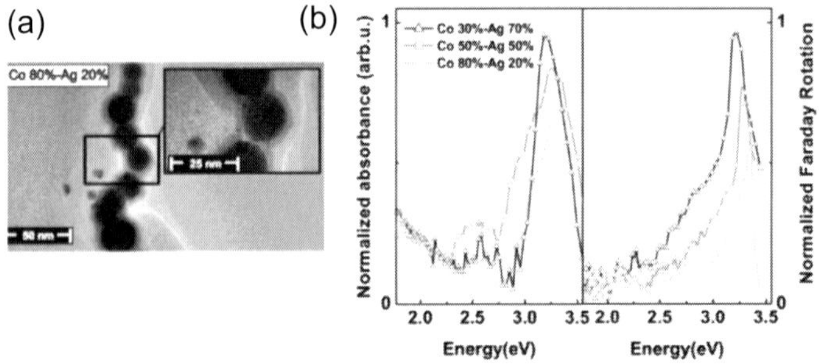

Figure 8.9 (a) TEM image of Co 80% Ag 20% nanoparticles shows a core–shell structure with an average diameter of 20 nm. (b) Absorbance and Faraday rotation spectra for Co–Ag nanoparticles with Co concentration ratios from 30% to 80%. Resonance-like peaks in the absorbance spectra are due to LSPR excitation, giving rise to a significant enhancement of the Faraday rotation. Adapted from Ref. [42].

pairs composed of a ferromagnetic and a noble metal [92], and core–shell nanoparticles [41, 42, 63, 74]. In particular, core–shell nanoparticles fabricated using colloidal chemical methods are of special interest since they exhibit a strong enhancement of their magneto-optical response, as predicted by Abe et al. [63, 93]. Indeed, an enhancement of the Faraday rotation has been detected experimentally in gold-coated magnetite nanoparticles [41], Ag–Co core–shell nanoparticles [42], and Ag–Fe core–shell nanoparticles [74]. Figure 8.9a shows a transmission electron microscopy (TEM) image for Co–Ag nanoparticles fabricated by aqueous synthesis [42]. This method allows fabrication of core–shell nanoparticles with different core and shell radii, while keeping the size of the nanoparticles constant at around 20 nm in diameter [94]. Transmission spectroscopy shows a resonance-like structure in the absorbance associated with LSP excitation for all the Co–Ag core–shell ratios (Fig. 8.9b, left). This excitation is highly dependent on the nanoparticles composition, and therefore the maximum values in the spectra appear blue-shifted from 3.2 to 3.35 eV as the Co concentration increases and so does the Co core diameter. Interestingly, a sharp increase of the Faraday rotation is obtained for those frequencies at which LSP modes are excited (Fig. 8.9b, right), which can be ascribed to the strongly enhanced electromagnetic

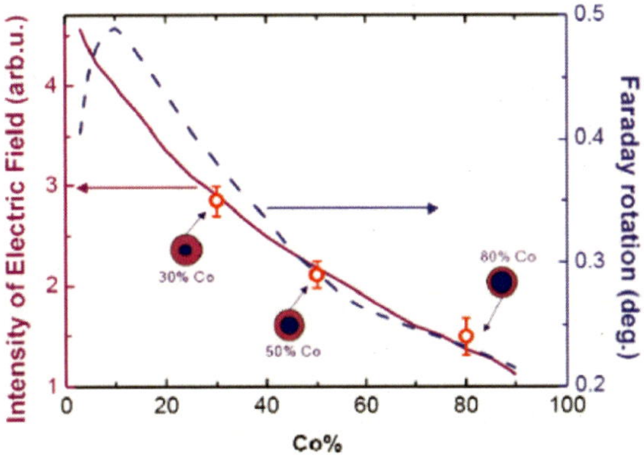

Figure 8.10 Calculated intensity of the electric field within the Co core in Co–Ag core–shell nanoparticles at resonance (continuous line) and maximum Faraday rotation of Co–Ag nanoparticles embedded in oil (n = 1.5018) (dashed line) as a function of the Co concentration. Calculations show excellent agreement with the experimental measurements (red dots). Adapted from Ref. [42].

fields inside the ferromagnetic core. To correlate the observed enhanced magneto-optical activity with the actual fields present at the magnetic core, the LSP-generated electromagnetic fields can be modeled by solving the boundary problem in Co–Ag core–shell nanoparticles that are exposed to an external static electric field in the quasistatic limit [63]. Indeed, a clear correlation between Faraday rotation and electromagnetic fields within the ferromagnetic core is found, as shown in Fig. 8.10. The calculated intensity of the electromagnetic field induced at the Co core (continuous line) grows with decreasing Co concentration in the nanoparticles, due to the higher absorption coefficient of Co, which hinders LSP excitation. Accordingly, the calculated maximum Faraday rotation (dashed line) increases with decreasing Co concentration in the nanoparticles, reaching its maximum at 10% Co concentration and decreasing for lower concentrations due to lack of sufficient magnetic material. It is worth noticing the very good agreement obtained between the experimental and calculated Faraday rotation values obtained at resonance (red dots) for the three Co:Ag concentration ratios studied. Thus, for the case of core–shell nanoparticles and similar nanostructures combining plasmonic and ferromagnetic materials in which LSPs can be excited, a clear correlation exists between the enhancement of electromagnetic fields and the increase of the magneto-optical response in an analogous way as for the case of thin films and multilayers sustaining SPP modes.

8.4 Conclusions

As it has been shown in this chapter, surface plasmonic resonances such as SPP and LSP modes have significantly attracted the attention of the scientific community due to their very intense associated electromagnetic fields, which are highly confined at the interface of the thin films and nanostructures supporting them. These plasmonic modes are able to concentrate and squeeze light well below the diffraction limit. Nevertheless, they are passive modes and cannot be externally controlled, which limits their use in practical applications. The adequate inclusion of ferromagnetic materials with high magneto-optical activity in these structures imparts them with a high sensitivity to external magnetic fields as, for example, in noble metal–ferromagnetic metal multilayers, nanocomposite thin films, multilayers on gratings, and core–shell nanoparticles. Additionally, a remarkable enhancement of the magneto-optical activity is observed due to the strong and highly localized electromagnetic fields generated by these plasmonic resonances. All these properties make these systems ideal candidates for applications such as, for example, all optical switches and modulators, plasmonic circuits, and biosensors.

Acknowledgments

We would like to acknowledge our collaborators L. Wang, K. Yang, J. R. Skuza, Z. Huba, K. J. Carroll, E. E. Carpenter, D. Gu, and M. Varela.

This work was supported by DARPA (grant # HR0011-07-1-0003, "Novel Sensors for Chemical and Biodefense").

References

1. H. Raether, *Surface Plasmons on Smooth and Rough Surfaces and on Gratings*, Berlin: Springer-Verlag, 1988.
2. H. A. Atwater, The promise of plasmonics, *Sci. Am.*, **April**, 56 (2007).
3. S. A. Maier, *Plasmonics Fundamentals and Applications.*
4. J. A. Schuller, E. S. Barnard, W. Cai, Y. C. Jun, J. S. White, and M. L. Brongersma, Plasmonics for extreme light concentration and manipulation, *Nat. Mater.*, **9**, 193–204 (2010).
5. N. Engheta, Circuits with light at nanoscales: optical nanocircuits inspired by metamaterials, *Science*, **317**, 1698–1702 (2007).
6. A. L. Falk, F. H. L. Koppens, C. L. Yu, K. Kang, N. de Leon Snapp, A. V. Akimov, M.-H. Jo, M. D. Lukin, and H. Park, Near-field electrical detection of optical plasmons and single-plasmon sources, *Nat. Phys.*, **5**, 475–479 (2009).
7. H. A. Atwater and A. Polman, Plasmonics for improved photovoltaic devices, *Nat. Mater.*, **9**, 205–213 (2010).
8. G. Garcia, R. Buonsanti, E. L. Runnerstrom, R. J. Mendelsberg, A. Llordes, A. Anders, T. J. Richardson, and D. J. Milliron, Dynamically modulating the surface plasmon resonance of doped semiconductor nanocrystals, *Nano Lett.*, **11**, 4415–4420 (2011).
9. J. Homola, *Surface Plasmon Resonance Based Sensors*, Berlin, 2006.
10. L. A. Loo, C. L. Hirsch, M. H. Lee, J. Barton, N. Halas, J. West, and R. Drezek, Nanoshell-enabled photonics-based imaging and therapy of cancer, *Technol. Cancer Res. Treat.*, **3**, 33–40 (2004).
11. N. Z. Abdel-Shahid and V. I. Pakhomov, On the study of the potential surface waves in plasma in the presence of an external magnetic field, *Plasma Phys.*, **12**, 55 (1970).
12. K. W. Chiu and J. J. Quinn, Magnetoplasma surface waves in metals, *Phys. Rev. B*, **5**, 4707–4709 (1972).
13. R. F. Wallis, J. J. Brion, E. Burstein, and A. Hartstein, Theory of surface polaritons in anisotropic dielectric media with application to surface magnetoplasmons in semiconductors, *Phys. Rev. B*, **9**, 3424 (1974).
14. P. E. Ferguson, O. M. Stafsudd, and R. F. Wallis, Surface magnetoplasma waves in nickel, *Phys. B+C*, **86–88**, 1403–1405 (1977a).
15. P. E. Ferguson, O. M. Stafsudd, and R. F. Wallis, Enhancement of the transverse Kerr magneto-optic effect by surface magnetoplasma waves, *Phys. B+C*, **89**, 91–94 (1977b).

16. J. Judy, J. Alstad, G. Bate, and J. Wiitala, Large longitudinal Kerr rotations and figures of merit in thin Iron films, *IEEE Trans. Magn.*, **4**, 401–405 (1968).
17. J. Judy, Variation of longitudinal kerr and faraday effects with angle of incidence in thin iron films, *IEEE Trans. Magn.*, **6**, 563–569 (1970).
18. R. K. Hickernell and D. Sarid, Long-range surface magnetoplasmons in thin nickel films, *Opt. Lett.*, **12**, 570–572 (1987).
19. B. Sepulveda, L. M. Lechuga, and G. Armelles, Magnetooptic effects in surface-plasmon-polaritons slab waveguides, *J. Lightwave Technol.*, **24**, 945–955 (2006).
20. J. J. Burke, G. I. Stegeman, and T. Tamir, Surface-polariton-like waves guided by thin, lossy metal films, *Phys. Rev. B*, **33**, 5186 (1986).
21. C. Hermann, V. A. Kosobukin, G. Lampel, J. Peretti, V. I. Safarov, and P. Bertrand, Surface-enhanced magneto-optics in metallic multilayer films, *Phys. Rev. B*, **64**, 235422 (2001).
22. J. B. Gonzalez-Diaz, A. Garcia-Martin, G. Armelles, J. M. Garcia-Martin, Clavero, C., A. Cebollada, R. A. Lukaszew, J. R. Skuza, D. P. Kumah, and R. Clarke, Surface-magnetoplasmon nonreciprocity effects in noble-metal/ferromagnetic heterostructures, *Phys. Rev. B*, **76**, 153402 (2007).
23. C. Clavero, K. Yang, J. R. Skuza, and R. A. Lukaszew, Magnetic field modulation of intense surface plasmon polaritons, *Opt. Express*, **18**, 7743–7752 (2010).
24. V. V. Temnov, G. Armelles, U. Woggon, D. Guzatov, A. Cebollada, A. Garcia-Martin, J. M. Garcia-Martin, T. Thomay, A. Leitenstorfer, and R. Bratschitsch, Active magneto-plasmonics in hybrid metal–ferromagnet structures, *Nat. Photon.*, **4**, 107 (2010).
25. G. Duchs, G. L. J. A. Rikken, T. Grenet, and P. Wyder, Magnetotransverse scattering of surface plasmon polaritons, *Phys. Rev. Lett.*, **87**, 127402 (2001).
26. D. M. Newman, M. L. Wears, and R. J. Matelon, Plasmon transport phenomena on a continuous ferromagnetic surface, *Europhys. Lett. (EPL)*, **68**, 692–698 (2004).
27. C. Clavero, K. Yang, J. R. Skuza, and R. A. Lukaszew, Magnetic-field modulation of surface plasmon polaritons on gratings, *Opt. Lett.*, **35**, 1557–1559 (2010).
28. M. S. Kushwaha and P. Halevi, Magnetoplasma modes in thin films in the Faraday configuration, *Phys. Rev. B*, **35**, 3879 (1987a).
29. M. S. Kushwaha and P. Halevi, Magnetoplasmons in thin films in the Voigt configuration, *Phys. Rev. B*, **36**, 5960 (1987b).

30. M. S. Kushwaha and P. Halevi, Magnetoplasmons in thin films in the perpendicular configuration, *Phys. Rev. B*, **38**, 12428 (1988).
31. M. Schubert, Polarization-dependent optical parameters of arbitrarily anisotropic homogeneous layered systems, *Phys. Rev. B*, **53**, 4265 (1996).
32. M. O. Vassell, Structure of optical guided modes in planar multilayers of optically anisotropic materials, *J. Opt. Soc. Am.*, **64**, 166–173 (1974).
33. G. Armelles, A. Cebollada, A. Garcia-Martin, J. M. Garcia-Martin, M. U. Gonzalez, J. B. Gonzalez-Diaz, E. Ferreiro-Vila, and J. F. Torrado, Magnetoplasmonic nanostructures: systems supporting both plasmonic and magnetic properties, *J. Opt. A: Pure Appl. Opt.*, **11**, 114023 (2009).
34. V. I. Safarov, V. A. Kosobukin, C. Hermann, G. Lampel, J. Peretti, and C. Marlière, Magneto-optical effects enhanced by surface plasmons in metallic multilayer films, *Phys. Rev. Lett.*, **73**, 3584 (1994).
35. Y. M. Strelniker and D. J. Bergman, Optical transmission through metal films with a subwavelength hole array in the presence of a magnetic field, *Phys. Rev. B*, **59**, R12763 (1999).
36. G. A. Wurtz, W. Hendren, R. Pollard, R. Atkinson, L. Le Guyader, A. Kirilyuk, T. Rasing, I. I. Smolyaninov, and A. V. Zayats, Controlling optical transmission through magneto-plasmonic crystals with an external magnetic field, *New J. Phys.*, **10**, 105012 (2008).
37. V. I. Belotelov, D. A. Bykov, L. L. Doskolovich, A. N. Kalish, and A. K. Zvezdin, Extraordinary transmission and giant magneto-optical transverse Kerr effect in plasmonic nanostructured films, *J. Opt. Soc. Am. B*, **26**, 1594–1598 (2009).
38. A. García-Martín, G. Armelles, and S. Pereira, Light transport in photonic crystals composed of magneto-optically active materials, *Phys. Rev. B*, **71**, 205116 (2005).
39. A. Battula, S. Chen, Y. Lu, R. J. Knize, and K. Reinhardt, Tuning the extraordinary optical transmission through subwavelength hole array by applying a magnetic field, *Opt. Lett.*, **32**, 2692–2694 (2007).
40. Y.-C. Lan, Y.-C. Chang, and P.-H. Lee, Manipulation of tunneling frequencies using magnetic fields for resonant tunneling effects of surface plasmons, *Appl. Phys. Lett.*, **90**, 171114 (2007).
41. P. K. Jain, Y. H. Xiao, R. Walsworth, and A. E. Cohen, Surface plasmon resonance enhanced magneto-optics (SuPREMO): Faraday rotation enhancement in gold-coated iron oxide nanocrystals, *Nano Lett.*, **9**, 1644–1650 (2009).

42. L. Wang, C. Clavero, Z. Huba, K. J. Carroll, E. E. Carpenter, D. Gu, and R. A. Lukaszew, Plasmonics and enhanced magneto-optics in core–shell Co–Ag nanoparticles, *Nano Lett.*, **11**, 1237–1240 (2011).
43. K. J. Chau, S. E. Irvine, and A. Y. Elezzabi, A gigahertz surface magneto-plasmon optical modulator, *IEEE J. Quantum Electron.*, **40**, 571–579 (2004).
44. T. Nikolajsen, K. Leosson, and S. I. Bozhevolnyi, Surface plasmon polariton based modulators and switches operating at telecom wavelengths, *Appl. Phys. Lett.*, **85**, 5833–5835 (2004).
45. J. B. Khurgin, Optical isolating action in surface plasmon polaritons, *Appl. Phys. Lett.*, **89**, 251115 (2006).
46. D. K. Gramotnev and S. I. Bozhevolnyi, Plasmonics beyond the diffraction limit, *Nat. Photon*, **4**, 83–91 (2010).
47. B. Sepúlveda, A. Calle, L. M. Lechuga, and G. Armelles, Highly sensitive detection of biomolecules with the magneto-optic surface-plasmon-resonance sensor, *Opt. Lett.*, **31**, 1085–1087 (2006).
48. K. Yang, C. Clavero, J. R. Skuza, M. Varela, and R. A. Lukaszew, Surface plasmon resonance and magneto-optical enhancement on Au–Co nanocomposite thin films, *J. Appl. Phys.*, **107**, 103924–103925 (2010).
49. E. Hutter and J. H. Fendler, Exploitation of localized surface plasmon resonance, *Adv. Mater.*, **16**, 1685–1706 (2004).
50. M. Fleischmann, P. J. Hendra, and A. J. McQuillan, Raman spectra of pyridine adsorbed at a silver electrode, *Chem. Phys. Lett.*, **26**, 163–166 (1974).
51. K. A. Willets and R. P. Van Duyne, Localized surface plasmon resonance spectroscopy and sensing, *Annu. Rev. Phys. Chem.*, **58**, 267–297 (2007).
52. B. Sepúlveda, P. C. Angelomé, L. M. Lechuga, and L. M. Liz-Marzán, LSPR-based nanobiosensors, *Nano Today*, **4**, 244–251 (2009).
53. X. Huang, I. H. El-Sayed, W. Qian, and M. A. El-Sayed, Cancer cell imaging and photothermal therapy in the near-infrared region by using gold nanorods, *J. Am. Chem. Soc.*, **128**, 2115–2120 (2006).
54. J. Chen, P. Albella, Z. Pirzadeh, P. Alonso-González, F. Huth, S. Bonetti, V. Bonanni, J. Åkerman, J. Nogués, P. Vavassori, A. Dmitriev, J. Aizpurua, and R. Hillenbrand, Plasmonic nickel nanoantennas, *Small*, **7**, 2341–2347 (2011).
55. V. I. Belotelov, L. L. Doskolovich, and A. K. Zvezdin, Extraordinary magneto-optical effects and transmission through metal-dielectric plasmonic systems, *Phys. Rev. Lett.*, **98**, 077401 (2007).

56. V. I. Belotelov, I. A. Akimov, M. Pohl, V. A. Kotov, S. Kasture, A. S. Vengurlekar, A. V. Gopal, D. R. Yakovlev, A. K. Zvezdin, and M. Bayer, Enhanced magneto-optical effects in magnetoplasmonic crystals, *Nat. Nano*, **6**, 370–376 (2011).
57. J. B. González-Díaz, A. García-Martín, G. Armelles, D. Navas, M. Vázquez, K. Nielsch, R. B. Wehrspohn, and U. Gösele, Enhanced magneto-optics and size effects in ferromagnetic nanowire arrays, *Adv. Mater.*, **19**, 2643–2647 (2007).
58. C. Clavero, A. Cebollada, G. Armelles, Y. Huttel, J. Arbiol, F. Peiro, and A. Cornet, Size effects in the magneto-optical response of Co nanoparticles, *Phys. Rev. B*, **72**, 024441 (2005).
59. C. Clavero, B. Sepulveda, G. Armelles, Z. Konstantinovic, M. G. del Muro, A. Labarta, and X. Batlle, Size mediated control of the optical and magneto-optical properties of Co nanoparticles in ZrO2, *J. Appl. Phys.*, **100**, 074320 (2006).
60. C. Clavero, G. Armelles, J. Margueritat, J. Gonzalo, M. G. del Muro, A. Labarta, and X. Batlle, Interface effects in the magneto-optical properties of Co nanoparticles in dielectric matrix, *Appl. Phys. Lett.*, **90**, 182506 (2007).
61. C. Clavero, L. Martinez, A. Garcia-Martin, J. M. Garcia-Martin, Y. Huttel, N. D. Telling, G. van der Laan, A. Cebollada, and G. Armelles, Morphology and capping effects in the magnetic and magneto-optical properties of nanoparticulate Co films, *Phys. Rev. B*, **77**, 094417 (2008).
62. Y. Huttel, H. Gomez, C. Clavero, A. Cebollada, G. Armelles, E. Navarro, M. Ciria, L. Benito, J. I. Arnaudas, and A. J. Kellock, Cobalt nanoparticles deposited and embedded in AlN: magnetic, magneto-optical, and morphological properties, *J. Appl. Phys.*, **96**, 1666–1673 (2004).
63. M. Abe and T. Suwa, Surface plasma resonance and magneto-optical enhancement in composites containing multicore-shell structured nanoparticles, *Phys. Rev. B*, **70**, 235103 (2004).
64. S. Melle, J. L. Menendez, G. Armelles, D. Navas, M. Vazquez, K. Nielsch, R. B. Wehrspohn, and U. Gosele, Magneto-optical properties of nickel nanowire arrays, *Appl. Phys. Lett.*, **83**, 4547–4549 (2003).
65. S. Tomita, T. Kato, S. Tsunashima, S. Iwata, M. Fujii, and S. Hayashi, Magneto-optical Kerr effects of yttrium-iron garnet thin films incorporating gold nanoparticles, *Phys. Rev. Lett.*, **96**, 167402 (2006).
66. S. Ozaki, H. Kura, H. Maki, and T. Sato, Manipulation of Faraday rotation in Bi-substituted yttrium-iron garnet film using electromagnetic interaction between Au nanoparticles in two-dimensional array, *J. Appl. Phys.*, **106**, 123530 (2009).

67. S. U. Jen and K. C. Chen, Enhancement of polar Kerr effect by forming Au nanoparticles on Ni surface, *J. Appl. Phys.*, **97**, 10M311 (2005).

68. G. Shemer and G. Markovich, Enhancement of magneto-optical effects in magnetite nanocrystals near gold surfaces, *J. Phys. Chem. B*, **106**, 9195–9197 (2002).

69. J. B. Gonzalez-Diaz, A. Garcia-Martin, J. M. Garcia-Martin, A. Cebollada, G. Armelles, B. Sepulveda, Y. Alaverdyan, and M. Kall, Plasmonic Au/Co/Au nanosandwiches with enhanced magneto-optical activity, *Small*, **4**, 202–205 (2008).

70. G. X. Du, T. Mori, M. Suzuki, S. Saito, H. Fukuda, and M. Takahashi, Evidence of localized surface plasmon enhanced magneto-optical effect in nanodisk array, *Appl. Phys. Lett.*, **96**, 081915 (2010).

71. D. Meneses-Rodríguez, E. Ferreiro-Vila, P. Prieto, J. Anguita, M. U. González, J. M. García-Martín, A. Cebollada, A. García-Martín, and G. Armelles, Probing the electromagnetic field distribution within a metallic nanodisk, *Small*, **7**, 3317–3323 (2011).

72. J. B. Gonzalez-Diaz, B. Sepulveda, A. Garcia-Martin, and G. Armelles, Cobalt dependence of the magneto-optical response in magnetoplasmonic nanodisks, *Appl. Phys. Lett.*, **97**, 043114–043113 (2010).

73. J. C. Banthí, D. Meneses-Rodríguez, F. García, M. U. González, A. García-Martín, A. Cebollada, and G. Armelles, High magneto-optical activity and low optical losses in metal-dielectric Au/Co/Au–SiO2 magnetoplasmonic nanodisks, *Adv. Mater.*, **24**, OP36–OP41 (2012).

74. L. Wang, K. Yang, C. Clavero, A. J. Nelson, K. J. Carroll, E. E. Carpenter, and R. A. Lukaszew, Localized surface plasmon resonance enhanced magneto-optical activity in core-shell Fe–Ag nanoparticles, *J. Appl. Phys.*, **107**, 09B303–303 (2010).

75. G. Armelles, J. B. Gonzalez-Diaz, A. Garcia-Martin, J. M. Garcia-Martin, A. Cebollada, M. U. Gonzalez, S. Acimovic, J. Cesario, R. Quidant, and G. Badenes, Localized surface plasmon resonance effects on the magneto-optical activity of continuous Au/Co/Au trilayers, *Optics Express*, **16**, 16104–16112 (2008).

76. J. F. Torrado, J. B. González-Díaz, M. U. González, A. García-Martín, and G. Armelles, Magneto-optical effects in interacting localized and propagating surface plasmon modes, *Opt. Express*, **18**, 15635–15642 (2010).

77. V. G. Kravets and A. S. Lapchuk, Enhancement of magneto-optical effects in magnetic nanoparticles near gold-dielectric surfaces, *Appl. Opt.*, **49**, 5013–5019 (2010).

78. N. Bonod, R. Reinisch, E. Popov, and M. Nevière, Optimization of surface-plasmon-enhanced magneto-optical effects, *J. Opt. Soc. Am. B*, **21**, 791–797 (2004).
79. L. Sapienza and D. Zerulla, Surface plasmon excitation on magnetoactive materials, *Phys. Rev. B*, **79**, 033407 (2009).
80. D. Regatos, B. Sepúlveda, D. Fariña, L. G. Carrascosa, and L. M. Lechuga, Suitable combination of noble/ferromagnetic metal multilayers for enhanced magneto-plasmonic biosensing, *Opt. Express*, **19**, 8336–8346 (2011).
81. E. Ferreiro-Vila, M. Iglesias, E. Paz, F. J. Palomares, F. Cebollada, J. M. González, G. Armelles, J. M. García-Martín, and A. Cebollada, Magneto-optical and magnetoplasmonic properties of epitaxial and polycrystalline Au/Fe/Au trilayers, *Phys. Rev. B*, **83**, 205120 (2011).
82. D. Regatos, D. Farina, A. Calle, A. Cebollada, B. Sepulveda, G. Armelles, and L. M. Lechuga, Au/Fe/Au multilayer transducers for magneto-optic surface plasmon resonance sensing, *J. Appl. Phys.*, **108**, 054502 (2010).
83. E. Ferreiro-Vila, J. B. González-Díaz, R. Fermento, M. U. González, A. García-Martín, J. M. García-Martín, A. Cebollada, G. Armelles, D. Meneses-Rodríguez, and E. M. Sandoval, Intertwined magneto-optical and plasmonic effects in Ag/Co/Ag layered structures, *Phys. Rev. B*, **80**, 125132 (2009).
84. C. Clavero, A. Cebollada, G. Armelles, and O. Fruchart, Growth mode, magnetic and magneto-optical properties of pulsed-laser-deposited Au/Co/Au(111) trilayers, *J. Magn. Magn. Mater.*, **322**, 647–652 (2010).
85. D. P. Kumah, A. Cebollada, C. Clavero, J. M. Garcia-Martin, J. R. Skuza, R. A. Lukaszew, and R. Clarke, Optimizing the planar structure of (111) Au/Co/Au trilayers, *J. Phys. D: Appl. Phys.*, **40**, 2699–2704 (2007).
86. P. Bertrand, C. Hermann, G. Lampel, J. Peretti, and V. I. Safarov, General analytical treatment of optics in layered structures: application to magneto-optics, *Phys. Rev. B*, **64**, 235421 (2001).
87. M. Ohring, *Materials Science of Thin Films*, San Diego, USA, 2002.
88. D. Stroud, Generalized effective-medium approach to the conductivity of an inhomogeneous material, *Phys. Rev. B*, **12**, 3368–3373 (1975).
89. A. A. Grunin, A. G. Zhdanov, A. A. Ezhov, E. A. Ganshina, and A. A. Fedyanin, Surface-plasmon-induced enhancement of magneto-optical Kerr effect in all-nickel subwavelength nanogratings, *Appl. Phys. Lett.*, **97**, 261908 (2010).

90. D. A. Bykov and L. L. Doskolovich, Magneto-optical resonances in periodic dielectric structures magnetized in plane, *J. Mod. Opt.*, **57**, 1611–1618 (2010).

91. C. d. J. Fernández et al., Coupling between magnetic and optical properties of stable Au–Fe solid solution nanoparticles, *Nanotechnology*, **21**, 165701 (2010).

92. Y. Li, Q. Zhang, A. V. Nurmikko, and S. Sun, Enhanced magnetooptical response in dumbbell-like Ag–CoFe2O4 nanoparticle pairs, *Nano Lett.*, **5**, 1689–1692 (2005).

93. M. Abe and J. Kuroda, Magneto-optical effects calculated for granular composites with magnetized nano-onions dispersed in matrixes, *J. Appl. Phys.*, **91**, 7305–7307 (2002).

94. K. J. Carroll, D. M. Hudgins, S. Spurgeon, K. M. Kemner, B. Mishra, M. I. Boyanov, L. W. Brown, M. L. Taheri, and E. E. Carpenter, One-pot aqueous synthesis of Fe and Ag core/shell nanoparticles, *Chem. Mater.*, **22**, 6291–6296 (2010).

Chapter 9

Understanding Magnetic Exchange in Molecule-Based Magnets from an Electronic Structure Point of View

Andrew P. Weber,[a] Paul M. Rulis,[a] Michelle M. Paquette,[a] Konstantin I. Pohkhodyna,[b] Saad Z. Janjua,[a] and Anthony N. Caruso[a]

[a]*Department of Physics and Astronomy, University of Missouri-Kansas City, Kansas City, MO 64110, USA*
[b]*Center for Nanoscale Science and Engineering, North Dakota State University, Fargo, ND 58102, USA*
carusoan@umkc.edu

9.1 Introduction

The vocation of science relates structure with function, identifying the links between physical forms and their properties. The science of solids is no exception; indeed, the field is home to a provocatively rich—perhaps infinite—variety of possibilities in the form of physiochemical structure. Despite this diversity, there are apparently only a few electronic phases of bulk solids, namely insulators, semiconductors, metals, and superconductors [1]. Meanwhile, the

Handbook of Nanomagnetism: Applications and Tools
Edited by Rosa A. Lukaszew

ISBN 978-981-4613-04-0 (Hardcover), 978-981-4613-05-7 (eBook)
www.panstanford.com

origin of magnetic properties concerns the formation of and communication between magnetic moments within a solid; this includes the magnitudes of these moments, the sign and strength of the coupling between them, and the range over which the resultant magnetic order extends. The origins of all of these phenomena are electronic at their foundation. We can agree that atomic moments may arise from the orbital motion of electrons; however, *in a solid*, it happens that the orbital contribution is usually very small in comparison to that of the intrinsic, spin magnetic moment of unpaired electrons [2]. It is the case that the spin-dependent interaction of electrons constitutes the primary basis for describing magnetism in a solid—the interactions require descriptions of electronic overlap within the context of both real space and binding energy space.

The principle means of spin coupling is the *exchange interaction* that arises as a result of the quantum mechanical constraints on the phase of a wave function when two component particles are interchanged. This interaction tends to only becomes significant when electrons are present within the same region of space and exist, or hybridize to exist, at the same binding energy. The electron–electron Coulomb repulsion and the Fermi statistics of electrons give rise to a spin-dependent energy representation of the solid's electronic structure. The effect of exchange is to couple the spins of electrons into parallel (ferromagnetic exchange) or antiparallel (antiferromagnetic exchange) alignment with one another. This decidedly binary result is an exhibition of quantum mechanics in nature. Within every magnetic solid, an intricate coordinated motion of polarized charges takes place. Like a ballet, the charges move within the solid that contains them with a tendency to avoid collisions (due to electron–electron Coulomb repulsion). They form a perplexing antisymmetry such that if the motion of one charge is spatially distinct from another, their spin polarizations are the same, but where motions overlap, the spin directions are opposite. Thus, electrons of like spin tend to avoid each other while electrons of opposite spin can be ambivalent. A final subtlety that keeps with the analogy to ballet is that the electrons, like dancers, prefer to maintain consistency in their direction of spin. If one dancer spins clockwise, the other will do the same as much as possible until no room remains on the stage. Then, the dancer will spin the other way and close ranks with another dancer. The intricate splendor of the many-electron structure only sets the stage as a medium to be rearranged by nuclear Coulomb energy, the dogged protagonist, who attempts to keep the charges localized.

If nuclear Coulomb energy succeeds in localizing the charges, as may happen in one way or another for good insulating materials, the phenomenon of exchange is often neatly summarized in energetic terms by the Heisenberg–Dirac–Van–Vleck (HDVV) Hamiltonian [3] (Eq. 9.1).

$$\hat{\mathcal{H}}^{\text{HDVV}} = -\sum_{i,j} \mathcal{J}_{ij}\vec{S}_i \cdot \vec{S}_j \qquad (9.1)$$

The Hamiltonian gives the energy resulting from the sum of pair-wise interactions between the spins $\vec{S}$ at sites i, j scaled by the *effective exchange parameter* $\mathcal{J}_{ij}$. The exchange parameter for a given pair-wise interaction is defined as the energetic difference between high-spin (same orientation spin) and low-spin (oppositely oriented spin) configurations. Respecting the energetically favored configuration of each interaction, it follows that $\mathcal{J}_{ij} > 0$ characterizes ferromagnetic coupling while $\mathcal{J}_{ij} < 0$ characterizes antiferromagnetic coupling of neighboring spins. A vast amount of work continues to be done in forming models based on spin–spin interactions that can incorporate more complicated phenomena, such as spin-orbit coupling and crystal fields (i.e. magnetic anisotropy) [4–8] and spin–lattice coupling [9–12]. It is also common to characterize a material's magnetic behavior on the basis of its physiochemical structure (the type of atoms or ions and their relative spacing and symmetry) and a phenomenological model such as that of HDVV [13–16]. These approaches only deal with spin interactions and do not consider the electrons (or electronic states) with which the spins are associated explicitly. Alone, the models obtained cannot provide a *mechanistic* description of how electronic interactions yield the sign and strength of exchange. Moreover, any spin–spin interaction is a fictitious notion, because the repulsion and antisymmetry of many electrons is responsible for exchange coupling rather than the spins themselves. Although the physiochemical structure and phenomenological models can be used to abridge the spectacle of

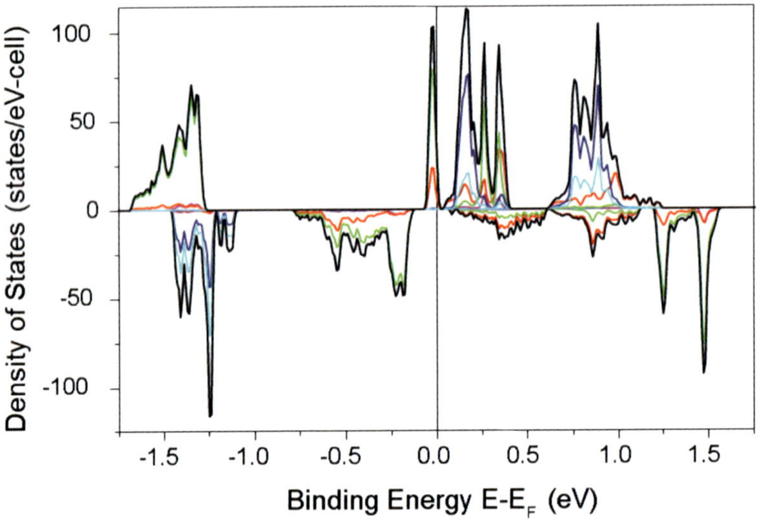

Figure 9.1 Calculated spin-resolved partial density of states of a heuristic metal–ligand coordination network. The upper panel is the spin majority, and the lower panel the spin minority. The black line represents the total density of states in each spin direction, while the green and blue represent the metal, and red, cyan, and magneta represent the ligand.

magnetic exchange, a complementary perspective is needed to gain the warmth of a front-row seat and hold the libretto in our hands. This is the *electronic structure* viewpoint. The electronic structure concerns the form and filling of electron quantum mechanical states, which constitutes a basis for understanding electron interactions and their effects. It holds the keys needed to unlock a deeper intuition for how magnetic exchange is controlled, and it may even lead to the prediction of subsequent properties (e.g., transition temperatures and their upper limits). As an example, consider the spin-resolved partial density of states (DOS) represented in Fig. 9.1 for a general metal–ligand solid-state system. In this case, strong hybridization exists between the traditionally localized metal states (green and blue) and delocalized ligand states (red, cyan, and magneta); it is this binding energy-dependent structure that allows one to visualize possible magnetic communication/exchange pathways.

Many questions pertaining to the coexistence of exchange-driven magnetism with specific chemical moieties and particular optical or electron transport properties can be put forward so as to target a novel application in technology or exploratory science. To these ends, molecule-based magnets (MBMs) are a promisingly adaptive class of solids poised to open new frontiers of exploration [17] for injectors/sensors of spin-polarized carriers, high-permeability materials, or low-density permanent magnets. MBMs are solids in which molecular subunits play a role in sustaining an exchange interaction that enables long-range magnetic ordering. The distinction of a subunit being molecular is made if it exists in an analogous, chemically stable form in solution or in the gas phase. In this way, MBMs are composed of molecular "building blocks," typically assembled at room temperature and often from solution, into a low-density framework, while conventional magnets are typically fabricated at high temperatures to yield a moderate-density lattice of atoms. This implies unique potential for MBMs. To begin with, the opportunity exists for solution-based synthetic chemistry to be applied in targeting compounds with specific physiochemical and subsequent electronic structures. This offers an enhanced prospect of "tenability," where it is expected that electronic and optical properties (e.g., band gap), in addition to structural and magnetic properties, can be modified through the systematic interchange of chemical moieties. Speaking to functionality in magnetic materials, photoinduced magnetism [18] and spin-polarized electron transport [19] have been suggested, predicted, and shown to exist in MBMs [20]. Organic MBMs are particularly promising key heterostructure layers in magnetoelectronic [21] applications, which rely on the selective transport of electrons with a particular spin orientation. By avoiding the incorporation of heavy elements, the strength of spin–orbit interactions (SOIs), which tend to rotate the electron spin dipole moments and encourage backscattering of electrons, is reduced in comparison to conventional inorganic materials proposed for the spintronics role (e.g., gallium arsenides) [22]. Moreover, the suggested tunability of electronic structure may allow for the design of so-called half-semiconductors or half-metals [23, 24].

Further contrasting MBMs with conventional magnets is that the *molecular* electronic structure is implicated in magnetic exchange interactions. For organic-based MBMs, the relevant molecular electronic structure we speak of is provided by the linear combinations of $2s$- and $2p$-orbitals, leading to σ- and π-orbital symmetries. Organic magnets are necessarily composed of radical (to include di- or multiradical) molecular species. Numerous organic radical magnets have been characterized (see, for example, reviews [25, 26]); however, we should note that their ordering temperatures are relatively low when compared to MBMs containing first-row transition metals (TMs). Organometallic coordination polymer magnets, such as antiferromagnetic copper phthalocyanine chains (CuPc) [27–29], ferrimagnetic metal tetracyanoethylenes (MTCNEs) [30–33], and the numerous Prussian blue analogues (PBAs) [34–36], possess unpaired electrons in $3d$-orbitals, which are implicated in magnetic exchange interactions. In these cases, exchange may occur directly between d-orbitals of adjacent atoms, as evident in CuPc chains [27], or indirectly through organic (ligand) molecular orbitals (MOs), as it happens in MTCNEs. There are also examples of $4f$-orbital magnetic exchange in lanthanide tetracyanoethylenes (LnTCNEs) [37], while prospects for $4d$- and $5d$-orbital molecular magnetism have been recently reviewed [38]. A variety of physical structures—quasi-1D chains, quasi-2D layers, and 3D covalent networks—can be obtained from MBMs with relatively similar chemical compositions. Figure 9.2 provides an illustration of this fact for compounds containing manganese coordination centers and tetracyanoethylene (TCNE) ligands.

Each of the three cases shown in Fig. 9.2 yields a progressively increased exchange strength and transition temperature (Fig. 9.2A$\rightarrow$9.2B$\rightarrow$9.2C) [39–41]; from a cursory perspective, the differences are related to the Mn oxidization state (i.e., number of unpaired electrons) and difference in the Mn-to-TCNE ratio, given that the relative coordination number is different. However, deeper inspection reveals that the stronger interactions are crystal field, orbital hybridization, and charge transfer dependent, all of which are quantities described by a combination of the local physical and electronic structures.

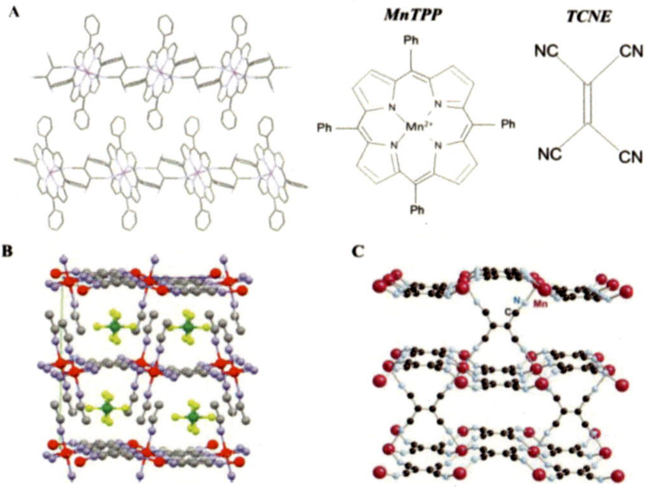

Figure 9.2 (Left A) Wireframe model of quasi-1D-structured $[Mn^{III}TPP]^+$ $(TCNE)^{\bullet-}$ [39]. Carbon atoms are marked by gray, nitrogen by blue, and manganese by purple. Hydrogen atoms are omitted for clarity. (Right A) Chemical structure of manganese tetraphenylporphyrin (MnTPP) and TCNE molecular units. (B) Ball-and-stick model view of three 2D layers of $[Mn^{II}(TCNE)^{\bullet-}(NCMe)_2]^+[SbF_6]^-$ (methyl hydrogen atoms omitted). Carbons atoms are marked by gray-, nitrogen by blue-, and manganese by red-colored balls, while antimony and fluorine atoms are marked by green and yellow, respectively. Adapted from Ref. [40] with permission. (C) Ball-and-stick model of 3D-covalent-network-structured $[Mn^{II}(TCNE)_{3/2}]^{+1/2}$ $[(I_3)_{1/2}]^{-1/2}$ [41] with iodate anions omitted for clarity. Carbon atoms are marked by black, nitrogen by blue, and manganese by magenta.

An additional class of MBMs consists of molecules containing metal ion clusters. Some of these molecular building blocks (e.g., $[Mn_{12}O_{12}(CH_3COO)_{16}(H_2O)_{24}] \cdot 2CH_3COOH \cdot 4H_2O$ (Mn12-ac)) are of fundamental interest on their own because they are superpara-magnets with strong magnetic bistability. That is, the molecular magnetization is anisotropic with only two stable orientations. The molecule tunnels from one magnetized state to the other at a rate determined by the temperature [42]. Although Mn12-ac does form magnetically ordered 3D crystalline structures [43], it is classified as one of many single-molecule magnets (SMMs) or "molecular nanomagnets," which constitute a field of study in and of themselves (see, for example, Refs. [44–47]). In spite of the intense interest SMMs have generated in the nanomagnetism community, their solids do not exhibit long-range magnetic order arising from intermolecular exchange interactions and are consequently not focused upon in this review. Despite the narrowing of scope, the concepts and methods discussed here that strive to understand magnetic exchange in MBMs should also be applicable to SMMs and inorganic magnetic systems in general.

9.2 Basic Views of Electronic Structure and Electron Interactions

9.2.1 *Competing Energies and Particle-Wave Structure*

Two primary interactions affect the orientation of an electron's spin, Larmor interaction and spin-dependent Coulomb repulsion. The former works by way of coupling to the spin's magnetic moment through an effective magnetic field, which is the case for SOIs (see, for example, Chapter 2 of Fazekas [2]) and the magnetostatic coupling of electron dipole moments. Although it would seem that electron dipole–dipole coupling could be responsible for long-range magnetic order, its relevance is a question of relative energies, namely thermal energy *versus* interaction energy. A generous estimate for the energetic strength of dipole–dipole interaction in a solid is typically on the order of 0.001 eV, in comparison with typical intra-atomic exchange energies of $\sim$10 eV and interatomic exchange energies of $\sim$0.1 eV. At 10 K, the thermal energy $k_B T$, where k_B is Boltzmann's constant is already above the dipole–dipole interaction energy (i.e., thermal disorder overcomes the order brought on by dipolar coupling very early on). Following the story of competing energies in a solid can lead to an appropriate description of its electronic and magnetic behavior as it helps identify the leading forces felt by electrons. Although the central theme of describing magnetic systems from a "bottom up" approach begins with understanding how electrons interact with one another in a spin-dependent fashion, it is equally important to identify the forces on electrons in a solid that act independently of spin. Otherwise a spin interaction model would lead to an entirely appropriate description.

Consider the familiar particle-in-a-box (PIB) problem, with particles confined to a region of zero potential energy by walls that give an infinite potential energy. The Schrödinger wave equation (SWE) is solved to yield wave functions that are zero-valued at the edges of the box. The solution in one dimension for a box of length a along a direction x is the set of sinusoidal particle wave functions. It follows that the energy of each independent, single-particle state is just its kinetic energy, $\varepsilon_n = \frac{\hbar^2 k_n^2}{2m}$. The momentum of the nth state is given by $\pm\hbar k_n$, which is inversely proportional to the length of the 1D box. In 1949, Kuhn [48] put forward a model based on PIB analysis to describe the optical absorption spectra (namely, the optical band gap) of organic dyes with conjugated π-orbitals, while Sommerfield developed a PIB-inspired model for electrons in metals. These very simple models, which captured the interplay between boundaries imposed by potential energy and the filling of electron states from low to high kinetic energies, had moderate success for their time and can be seen as precursors to what are now the ubiquitous MO theory of chemistry and the band theory of solid-state physics. In both MO theory and band theory, valence electron wave functions are obtained by treating ions (nuclei with core electrons) as attractive potentials that scatter otherwise freely moving electrons. The effect of the ions is to cause discontinuities in electron wave functions, leading to allowed and disallowed momenta and energies, just as in the PIB case. The only difference between band theory and MO theory is in the boundary conditions the wave functions must satisfy.

The prototypical model of chemical bonding is the linear combination of atomic orbitals (LCAO) approach to MO theory. Atomic wave functions φ_A and φ_B on neighboring atoms A and B within a molecule are added in phase to yield a bonding MO

$$\psi_\mathrm{bond} = \varphi_\mathrm{A} + \varphi_\mathrm{B}$$

of even symmetry and out of phase to yield an antibonding MO

$$\psi_\mathrm{anti} = \varphi_\mathrm{A} - \varphi_\mathrm{B}$$

of odd symmetry in the space between A and B. The structures of the even and odd wave functions show that the antibonding orbital has a nodal surface of zero electron density between the two atoms, whereas the bonding orbital does not. The odd wave function will have higher kinetic energy; hence the bonding orbital is filled before an antibonding orbital. This fact alone does not explain the stability of a bond that contains two electrons that necessarily repel one another. The Pauli exclusion principle provides the natural explanation for the stability of a chemical bond; antiparallel coupling of electron spins allows their paths to overlap in space. The bonding electrons can then move away (delocalize) from their "home" atoms into the region between the atoms, reducing their kinetic energies yet even more.

In the case that an antibonding orbital becomes filled (while there are also other bonds present that stabilize the molecule), the electrons within that orbital must still remain antiparallel, but it is for a reason that cannot be reconciled by atomic orbital overlap, because there is none to speak of by virtue of the structure of the antibond. The subtlety lies in the fact that each electron within an MO does not "belong" to a particular atomic orbital. Recall the view of electrons scattering from ion to ion with particular, allowed momenta. It may happen that electrons end up in the same place at the same time; hence it is required that two electrons within a single MO must have their spins coupled antiparallel. This much is true for both bonding and antibonding orbitals. However, antibonding orbitals feature electron densities that are more localized to the individual atoms, indicating that the probability of two electrons occupying the same atomic orbital of the MO is enhanced when compared with the bonding configuration. It happens that including electron interactions into the model causes the energy difference between the bonding and antibonding orbitals to increase. Certainly, there is always an energetic cost associated with putting two electrons in the same localized space. Considering the addition of an electron to an already half-filled orbital site, we call this the *on-site Coulomb repulsion U*. If *U* presents a strong energy barrier for additional occupation, then electrons will tend to remain localized to respective atoms This idea becomes more important when bonding/antibonding interactions occur between multiply spin-bearing sites, which is frequently spoken of in the context of models proposed by Mott [49] and Hubbard [50] that were meant to explain both the antiferromagnetism and insulating behavior of TM oxides.

9.2.2 *Ligand Fields and Magnetic Exchange in Organometallic MBMs*

In a TM–organic coordination complex (e.g., $Cr(CO)_6$), the electrostatic charge of ligand electron density causes a nontrivial perturbation (electrostatic crystal field) to the local electronic structure of the coordinated TM ion. The crystal field causes the fivefold *d*-orbital degeneracy to be lifted according to the orbitals' symmetries relative to that of the crystal field (single-ion anisotropy) and the quenching of *d*-orbital angular momentum. An octahedral crystal field (O_h) causes a splitting of the *d*-shell into twofold degenerate e_g and threefold degenerate t_{2g} levels separated by a crystal field energy Δ_{CF} (see Fig. 9.3). The d_z^2- and d_x^2–y^2-orbital densities are focused along the axes at which the ligands coordinate. In contrast, the t_{2g}-orbitals are directed toward the relatively empty space between the ligands, and therefore, their interaction with the ligands' electrostatic potential is weaker. As a result of the desire to avoid strong Coulomb repulsion between the metal and ligand electrons, the e_g levels' energy will shift to a higher energy than the t_{2g} levels. The ground state for the e_g and t_{2g} levels is generally derived according to Hund's rule (i.e., total spin S should be large as possible to minimize the on-site Coulomb repulsion). For instance, the V^{2+} (d^3) ion has a spin S equal to 3/2 for three electrons with parallel spins residing on three degenerate t_{2g} levels. For *d*-orbital occupations greater than three electrons, the octahedral crystal field energy will play a competing role with Coulomb energy summarized by Hund's rule. Thus, when Δ_{CF} is sufficiently large, the electron filling no longer follows Hund's rule and the low spin state arises.

Taking the step from a crystal field theory (CFT) (where the metal 3*d*-orbitals simply experience Coulomb repulsion from point charge-like ligands) to ligand field theory (LFT) requires considering that the metal 3*d*-orbitals overlap with the MOs of the ligands, forming new bonding and antibonding MOs. The bonding MOs have a larger contribution from the ligands, because the ligand orbitals are lower in energy, while the antibonding MOs have a large metal contribution, because the metal orbitals are higher in energy. The three TM 3*d* t_{2g}-orbitals generally remain nonbonded since there is no orbital of the same symmetry in the ligand orbitals. These three

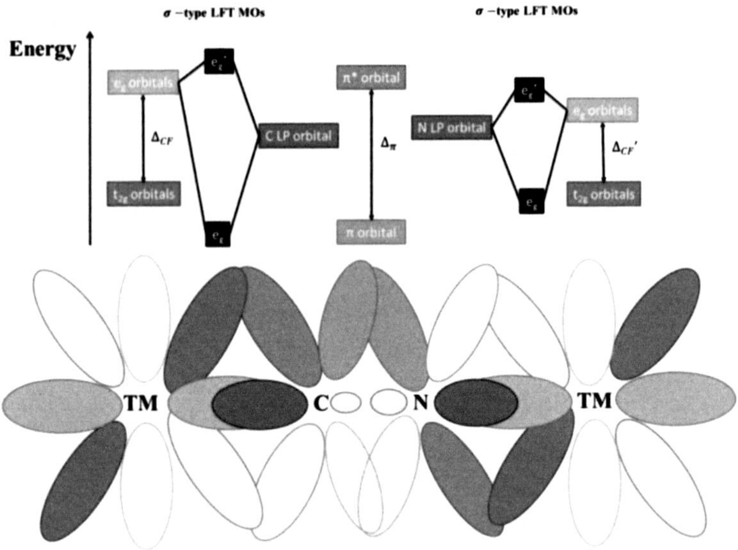

Figure 9.3 A cartoon of a possible ligand/metal orbital overlaps in a cyano-bridged metallate, with color-coordinated ovals representing orbital densities. The top of the figure shows a proposed energy diagram for the molecular and TM electronic states involved.

TM 3d-orbitals with xy, xz, and yz symmetry are then nearly the same in energy as in the isolated TM ion. However, in some cases there can be a weak TM t_{2g} and ligand π^* antibonding MO overlap, which can result in so-called π backbonding usually accompanied by metal-to-ligand charge transfer (MLCT). Meanwhile, the e_g-orbitals of the TM ion do overlap with p-type, or lone pair (LP), orbitals of the ligand to form a set of bonding orbitals, denoted as e_g, and antibonding orbitals, denoted as e_g*. The splitting between the e_g and e_g* ligand field MOs increases with the ligand/metal orbital overlap. Because the bonding orbitals are cylindrically symmetric on the coordination axis, they are referred to as σ-type MOs.

Understanding the interactions between ligand and TM orbitals requires both a spatial picture and an energetic picture of the electronic structure. As an example, Fig. 9.3 illustrates a real space and binding energy space picture of a cyano-bridge of a hexacyanometallate complex.

The σ-type bonding is stronger and Δ_{CF} is larger for coordination at the carbon side of the bridging molecule because carbon is more electropositive than nitrogen, and its LP orbital will therefore penetrate the metal e_g-orbital to a greater extent, creating a greater splitting in energy between the e_g* and e_g LFT MOs. One can imagine that the cyanide π^* MO could hybridize (backbond) with metal t_{2g}-orbitals on opposite sides, since a weak overlap appears possible. This could form an electronic hybrid state encompassing both TMs in the picture, and hence, spins resident on both sites may become exchange-coupled. However, the CN π^* is generally too high in energy for a large backbonding to occur that can explain magnetism in a cyano-bridged metallate. Many case studies of the type of complexation shown in Fig. 9.3 have been undertaken in the context of PBA MBMs, which display a variety of electronic structure, distortive electron–lattice interactions, and magnetic behavior depending on the chemical composition of the lattice (cyano-bridged metallates) and its interstitial counter ions (which affect the mean valence of ions and molecules in the unit cell) [36].

The HOMO/LUMO gap Δ_π of the bridging molecule can be tuned by replacing CN with a longer molecule containing conjugated π bonds. On account of the increased size, one can reason (even from a simple PIB model) that of the π MOs energies and Δ_π will both be reduced. Energetically aligning the π^* state with the t_{2g} states opens the energetic possibility that backbonding can occur, and play a substantial role in mediating an exchange interaction through the molecule's π^*-orbital, as Kortright et al. [51] proposed to explain magnetic and electronic structure properties of V-TCNE MBMs. Even in applying the same analysis (see Section 9.4) to the same system, a separate group, that of Fahlman and coworkers, arrived at a different conclusion that magnetic coupling in VTCNE was the result of charge-transfer interactions between MOs and the unoccupied e_g states of the V^{2+} coordination center.

Without a detailed knowledge of the energy distribution of electronic states, very little can be properly rationalized about magnetic coupling pathways. In later sections, we summarize this information in terms of the DOS relative to the point of zero-binding energy (the Fermi level E_F). Binding energy-dependent electronic structure can be probed through spectroscopic experimental methods (Section 9.4). Obtaining an accurate model based on the calculation of electronic structure for magnetic systems, however, is a formidable undertaking that begins with the problem of accounting for electron–electron interactions.

9.3 Formal Theories of Exchange–Correlation

9.3.1 *The Independent-Electron Approximation as a Starting Point*

A first attempt at modeling electronic structure can be made by considering only the nuclear Coulomb attraction, the electron kinetic energy, and by then filling the independent electron states according to the Pauli exclusion principle, while ignoring all Coulombic electron–electron interactions. This method can often be valid in systems with weakly interacting electrons because the electron kinetic energy and nuclear potential energy terms are the most dominant in determining the bound states when compared to electron–electron interaction energies. (Indeed, if this were not the case, there would be few bound states to speak of because electrons always repel one another.) Following the above approach is often termed an *independent-electron approximation*, as each electron state is expressed independently without considering the presence of other electrons except via the Pauli exclusion principle. To make such an approach realistic, however, the presence of core electrons about the nuclei should be taken into account by appropriately attenuating the electrostatic nuclear potentials in the problem, creating the so-called screened or effective nuclear potentials. It is assumed that removing the problem of interaction allows for exact solutions of the eigenstates $|\psi_{n0}\rangle$ of $\hat{\mathcal{H}}_0$, each with its own corresponding eigenenerg $y\varepsilon_{n0}$ to be found

$$\varepsilon_{n0}\,|\psi_{n0}\rangle = \hat{\mathcal{H}}_0\,|\psi_{n0}\rangle$$

Because spin-dependent interactions were not considered within $\hat{\mathcal{H}}_0$, each state is said to be twofold degenerate in energy with respect to spin orientation (when ignoring spin–orbit coupling). Therefore, we can include spin by simply multiplying the spin states with the electron states as $\left|\psi_{n,0}\right\rangle|\uparrow\rangle$ or $\left|\psi_{n,0}\right\rangle|\downarrow\rangle$. The same energies (ε_{n0}) will still be achieved.

The independent-electron treatment cannot provide a full account of the electron configurations of molecules or atoms. The crucial factor left missing is the possibility that some states may be close enough (or equal) in energy such that electron–electron repulsion plays a role in determining the spin configuration of the system. In the atomic case, a student of physics or chemistry will recall Hund's rules, which provide a systematic guide to the filling of atomic orbitals with the correct relative orientations of electron spins. Hund's rules are nothing more than a phenomenological means of accounting for electron–electron interactions and spin–orbit coupling. An adequate description of electronic structure must account for electron repulsions, which are treated in different ways at different levels of physical theory. To obtain a model of how repulsive forces affect the configuration of atoms, molecules, or solids, it is common to start from an independent-electron theory solution to the problem and then add the sum of pair-wise electron–electron Coulomb interactions:

$$\hat{V}_{e-e} = \sum_{i,\,j} \frac{e^2}{\left|\vec{r}_i - \vec{r}_j\right|}$$

as a perturbation. Note that this will raise the energy of all electrons in the model. We can choose to consider interactions only between electrons whose independent-electron theory energies are close or equivalent, for it is here where forces besides electron repulsion have little or nothing, respectively, to say concerning the need for specific spin configurations. However, evaluating the energy of electron interactions is not a straightforward task.

There are subtle, though important, differences for thinking in terms of one electron interacting with another as opposed to many electrons interacting as a whole. A point of conceptual conflict often arises when studying quantum mechanical many-body systems, such as a solid. One vision is that the elementary components of matter are individual particles acting on their own, according to the complex environment in which they find themselves. This would appear much like people walking past each other in a crowd and where the crowd's overall behavior is derived from the discrete choices of individual people. Continuing the analogy, the formulation of quantum mechanics, however, regards the crowd itself as a single entity with certain overarching characteristics imposed upon its subcomponents. The conventional view of individual interacting particles still has merit because it offers a tractable way to approximate the many-body complications that arise from the quantum mechanical theory. The formalism is thus quantum mechanical, but the viewpoint when describing a system is often semiclassical. However, because the overarching nature of the single system wave function should always be kept in mind, it is emphasized in this section through the frequent (superfluous) description of electrons as "component" electrons.

The essential ingredient for understanding the role of component electrons in a single overarching many-body system (e.g., a free electron gas or condensed matter) is that the electrons are indistinguishable particles that cannot occupy the same space with the same set of quantum numbers, as from the Pauli exclusion principle. Implicit in this statement is the fact that the electrons are all mutually affected by each other such that the motion of any given component electron depends on the positions of every other electron at a given point in time. Additionally, it is important to note that while this effect is entirely independent of the classical Coulomb force that acts to repel electrons from each other it does have the overall net effect of reducing the probability that two component electrons will be found near each other. Thus, this effect acts in a similar repulsive way as the classical Coulomb force does but to a different, lesser, magnitude. Formulating methods to express these classical Coulomb and quantum mechanical characteristics in computationally tractable ways has taken the traditional divide-and-conquer approach whereby parts of the problem that can be easily addressed via one method or another are sectioned off and treated appropriately, while other (hopefully small) parts that do not lend themselves to simple treatment are either ignored or approximated with varying levels of sophistication (e.g., Jacob's ladder of density functionals [52] or the broad hierarchy of quantum chemical methods [53]).

Following this approach, the first division is made according to so-called exchange and correlation terms. Exchange is recognized as the portion of the electron–electron interaction that is derived from the quantum mechanical requirement that the many-electron wave function be antisymmetric with respect to interchange of any space or spin variable, as discussed in the preceding paragraph. Thus, the overall wave function must change sign when the space or spin coordinates of two component electrons are interchanged. Because all electrons have spin half the interchange can only be in the spatial coordinate. In the particular cases of atomic and molecular chemistry the spatial coordinate that most strongly affects the average proximity of electrons to each other is the *m* quantum number and thus the "exchange interaction" is manifest as Hund's Rule. In the Hartree–Fock method this issue can be incorporated explicitly and exactly into a condensed matter many-body system. As a point of semantics, the "exchange hole" that is often discussed is equivalent to the accumulated repulsive effect of the antisymmetry requirement that reduces the likelihood of finding another component electron in the neighborhood of a given component electron.

Correlation is most commonly recognized as the portion of the electron–electron interaction that remains when a simple approach that accounts for classical Coulomb electron–electron repulsion between the average spatial distributions of the electrons is already taken into account. In other words, the electrons are envisioned to be point particles that all mutually interact via a Coulomb force, but we choose to simplify the interaction by treating them according to their individual average charge distributions. In this way we can account for a substantial portion of their mutual interaction (specifically the long-range Coulombic aspect), but the short-range details that are smeared out by simply using *average* electron charge distributions remain unaccounted for and are collectively called "correlation." Using average distributions for the component electrons is clearly not the same thing as explicitly considering their mutually interacting particle nature at close distances (e.g., on-site and nearest neighbor). Because the use of average charge distributions tends to allow for overall smaller distances between electrons than would be possible (on average) were the electrons treated individually, the correlation term that remains to be accounted for will have a net repulsive effect.

This accumulated repulsive effect between all pairs of component electrons is equivalent to the often discussed "correlation hole."

The exchange and correlation contributions to the potential energy function are critical points of consideration in the development of improvements to electronic structure calculations. As such, a considerable collection of information has been published in diverse collections of media. A starting point sample includes textbooks [54–56], monographs [57], conference proceedings [52], journal articles [58, 59], Internet websites [60, 61], and the array of references contained therein.

9.3.2 *Theories for Evaluating Exchange and Correlation*

9.3.2.1 Extensions of the Hartree–Fock method

In the Hartree–Fock method the many-body wave function is expressed by associating a collection of atomic orbitals and spin directions with each electron such that the size of the collection of spin–orbital pairs is at least equal to the number of electrons. Then, the sets of spin orbitals are organized into the elements of a matrix with a row for each component electron and a column for each possible spin orbital. This representation is convenient because when the determinate of the matrix is taken it will produce a formula for the many-body wave function that is automatically antisymmetric with respect to the interchange of electron coordinates. This approach explicitly and exactly accounts for the exchange interaction in its entirety.

The Coulomb interaction in the Hartree–Fock method is treated much as was described for general understanding in the earlier paragraphs. The long-range Coulomb interactions between electrons are accounted for through calculation of interactions between the average charge distributions of pairs of electrons. Mathematically this is computed via integrations over the spin orbitals that describe the electron distributions. The remaining short-ranged correlated electron motion is ignored in the Hartree–Fock method resulting in an upper bound on the total energy of the system. This is an upper bound because the ignored correlation term would (if it were included) imply a tendency toward greater separation between component electrons than was otherwise calculated. If the electrons should tend toward greater separation than was computed so far, then the energy needed to bring the system together from infinity should be less than was computed to this point.

The neglect of short-range correlation in the Hartree–Fock method can be corrected for with a conceptually simple approach that is also computationally prohibitive. Instead of using a single determinant to define the system wave function, a linear combination of determinants is used where the distribution of electrons across all of the various possible spin orbitals is different for each determinant. The coefficients of the linear combination of determinants are varied until a minimum in total energy is found. In this way, given sufficient variational freedom in terms of the number of spin orbitals available and given a sufficient number of variations in terms of different configurations for the electrons, the tendency for component electrons to be more dispersed than would be found from a single determinant will be incorporated. Thus, the short-ranged correlation effect can be accounted for. The practical difficulty with this so-called configuration interaction approach is that the number of configurations increases exponentially with the number of electrons. This renders the technique all but impossible for practical use in large molecules or extended systems of any significant size. However, the technique is still used in many cases because of its high accuracy and theoretical rigor and it has been developed and extended in a variety of different ways that enhance its utility [62, 63].

9.3.2.2 Density functionals

Within density functional theory (DFT), the long-range Coulomb interaction is accounted for in much the same way as in Hartree–Fock theory. However, in DFT the exchange and short-ranged correlation terms are lumped together and computed in an approximate way (It is necessary to approximate the exchange term in DFT because the system wave function is not written in a properly antisymmetrized form). The method assumes that, because the exchange and correlation terms are short ranged, it should be possible to sample a sufficient number of points in space and accumulate the local exchange and correlation interactions from each point into a correct total exchange and correlation interaction. If a method (functional) could be devised that predicts the contribution to the exchange and correlation energies only on the basis of local information from each point, then the complicated local electron–electron interactions would be fully incorporated into the total energy calculation.

The first application of this approach made use of auxiliary highly accurate quantum Monte Carlo (QMC) calculations [64] of the exchange and correlation effects in a uniform free-electron gas. QMC results from different gas densities were then fitted to relatively simple functions so that when the time came to sample a given point from the actual system of interest all that was needed was the density at the given point. In other words, the density from a point in the system of interest is given to the fitted function that subsequently returns the exchange and correlation energy contributions. The accumulated effect of all sample points provides a good measure of the total exchange and correlation effect. The local density approximation (LDA) for the exchange and correlation works remarkably well, even though the density in the system of interest (e.g., a periodic solid) is typically highly non-uniform. The unexpected success is due to a fortuitous cancellation of errors [59]. The next logical step beyond the LDA is the local spin density approximation (LSDA), which follows a similar conceptual approach except that the up and down spin densities are sampled instead of the total density. A logical step beyond the LSDA is the so-called generalized gradient approximation (GGA), where the gradient of the charge density through the sampled region of space is also considered. For each of these levels of approximation there exist a wide variety of different functionals that have been developed through the past decades [64–75]. However, all of these approaches include some deficiencies that limit their accuracy and thus applicability to certain problems.

An example deficiency in DFT methods becomes apparent when it is recalled that the effect we are seeking to quantify (exchange and correlation) is an effect on a given electron due only to the other electrons (i.e., the given electron does not respond to itself) but that the charge density we are sampling is the charge density due to *all* electrons. Thus, each of the above functionals introduces a spurious self-interaction. As another example, the assumption that the exchange and correlation of the uniform electron gas can be used to predict the exchange and correlation of a nonuniform charge distribution is faulty. Indeed, for systems where the electron density is large, highly nonuniform, and sensitive to small variations in neighboring distributions (e.g., in the region of partially occupied 3-, 4-, and $4f$ electron orbitals) the assumption breaks down and the electrons are termed "highly correlated," meaning that they experience a stronger correlation effect than a uniform electron gas density of similar density would.

The last example deficiency is likely the most important. The fundamental challenge with the DFT approach outlined in this section is that there is no systematic way to improve the exchange–correlation calculation. Improvement will not come simply by adding more terms to a list or by obtaining a more accurate uniform electron gas description. Rather, it must come from a more complete accounting of the observed deficiencies or from wholly new ways of modeling the sources of electron–electron interaction. These deficiencies have spurred on a search for better functionals or better overall methods of accounting for the exchange and correlation effects. However, it should be noted for a wide array of materials and material properties the above functionals provide very useful results with only limited computational effort. In the case of LDA, the solid-state band gap of insulators is typically underestimated by 20%–30%, the lattice parameters of simple crystals are often underestimated by $\sim$3%, and binding energies can be in error by several electron volts, but these errors are systematic and thus the results can still be used as viable predictions. GGA tends to improve upon this variety of LDA results but sometimes in an inconsistent way so that it becomes difficult to identify trends. For example, the lattice parameters may be significantly corrected in numerous cases, but then sometimes (e.g., for copper) they are overcorrected by the same degree of error that the LDA approach originally had [76].

9.3.2.3 Hybrid functionals

In principle, the approach used by the Hartree–Fock method to incorporate the exchange effect exactly could also be used within DFT. The orbitals used in a DFT calculation may also be arranged according to the requirement of antisymmetric exchange. The challenge though is that the correlation effect remains unaccounted for, and to include it, the existing exchange–correlation functionals would need to be decomposed into respective explicit exchange and correlation components. The exact decomposition depends on the coupling between the two [77]. The inclusion of a portion of the exact exchange and a weighted use of the typical exchange–correlation functionals in different proportions represents the development of so-called *hybrid functionals*. The degree of exact exchange and weighted use of LDA/GGA for a particular hybrid functional is generally a function of the types of systems (e.g., organic molecules) or properties (e.g., energy of formation) for which the functional has been developed. Thus, as may be expected, this approach is not as general as one would hope in that the various hybrid functionals that have been developed [73, 78–81] are not necessarily equally applicable to any type of material (e.g., metals) or calculable property (e.g., prediction of bond lengths) [59]. While understandable, the major problem remains that there is no systematic way to improve the exchange–correlation calculation in a way that is applicable to all materials. It is therefore considered good practice to compare the results obtained from different hybrid methods when investigating a specific system. As an example, in the study of M=Mn and V in the M[TCNE]s, Erdin [82] applied four different hybrid functionals for comparative study, finding slightly different results for each. Although each functional method seemed to reproduce within an order of magnitude the difference in high-spin and low-spin states that would correspond to the Curie temperatures of the ferrimagnets, we should note that the DOS ultimately proposed based in study was inconsistent with experimental results obtained from photoemission [83]. This could be an indication that correlation effects stronger than those that can be accounted for in LSDA or Hartree–Fock may play a critical role in the binding energy-dependent electronic structure of M[TCNE]s.

9.3.2.4 DFT+U

To account for strong, short-range correlations that concern particular, localized sites of electron occupation, the on-site Coulomb repulsion U can be assigned for particular orbitals, such as the $3d$-orbitals of a TM ion. Implementations of the Hubbard U in DFT are known as DFT+U theories [84–86]. The calculation requires that a set of local orbitals be defined, often the so-called projector-augmented waves (PAWs) [87]. The U is applied as a spin-dependent energy for an electron occupying a particular orbital site. The numerical value of U can be either estimated from empirical results [84] or calculated from first principles using one of several methods involving the all-electron calculation of the ion within a crystal field [88–90]. The energetically reshaped atomic orbital states are then incorporated into the calculation for the entire system. While it is common for DFT+U to be applied to strongly correlated electron systems, its application to MBM systems appears to be limited to a relatively low number of studies (e.g., [91, 92]) when compared to the application of hybrid functional and other methods. Perhaps this ought not to be the case. For example, Oppeneer and coworkers have systematically shown that a GGA+U formalism is needed to describe the local electronic structure ligated and surface-interfaced of TM-centered porphyrins [93–95].

9.4 Spectroscopic Probes of Electronic Structure: Application to Metal Tetracyanoethylenes

For magnetic systems, the intent of most electronic structure models is to elucidate the fundamental interactions that give rise to measureable magnetic properties, having taken some input of the physiochemical structure as a starting point. With this aim in mind, it would appear that the most relevant quantities extracted from such models are the exchange parameters $\mathcal{J}_{ij}$ that characterize the effective intersite exchange energies within the solid. The problem would then be to establish an appropriate spin-Hamiltonian describing the difference in interaction energy between high-spin and low-spin states, perhaps followed by a statistical treatment to yield the behavior of a system's magnetization with temperature or simply estimate the transition temperature. Each step in this process, especially beginning with the electronic structure calculation, is rarely straightforward or rigorously justifiable (see [96–98]). It is therefore desirable that models be tested across multiple parameters of experimental measurement in addition to the standard characterization of thermal and magnetic properties. Our conviction toward this end is so strong that we have included a separate section specifically intended to highlight the relation of photon and electron spectroscopy experiments to the spin-polarized electronic structure.

While values proportional to magnetic moment, exchange strength, exchange type and physical structure are quantitatively extracted for most MBMic materials, there are few studies that quantitatively probe the binding energy-dependent or state-specific electronic structure. Here we review and discuss the methods and results of some electron and photon spectroscopies, via a case study of $M[TCNE]_x$ systems, in an effort to understand the limitations and historic validations of electronic structure theory/models as well as reach out to the chemistry-dominated MBM community concerning lesser known techniques. The ultimate goal is to directly measure the energy and intensity of the occupied and unoccupied spin-polarized partial DOS and dispersions thereof relative to the Fermi edge. Because of limitations in the source (e.g., incident light polarization), sample (e.g., polycrystalline character, photodegradation rate, oxygen sensitivity), and/or sense (e.g., low signal-to-noise analyzer or quantum mechanical restrictions due to a forbidden transition) one must cobble together results in light of the models proposed, from a number of imperfect and complementary techniques to stitch together the information needed.

9.4.1 *Photoemission Spectroscopy*

Photoemission spectroscopy provides a measure of quantities proportional to the occupied DOS relative to the Fermi level or 0 eV binding energy. In the photoemission experiment a photon of well-defined energy $h\nu$ is incident on the sample surface; in

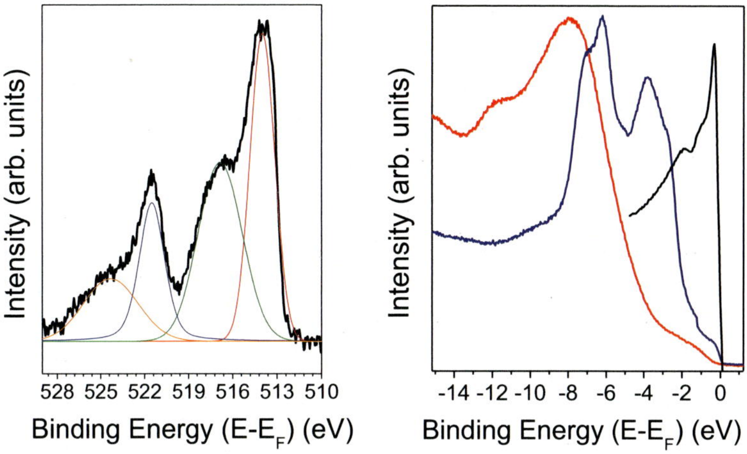

Figure 9.4 (Left) X-ray photoemission of the V[TCNE]$_x$ $2p$ core level showing the V $2p_{3/2}$ and V$2p_{1/2}$ core levels in a mixed V(II) and V(III) state. Adapted from Ref. [99] (Right) Generic valence band photoemission of Fe-TCNE (red), clean Au (blue) and clean Co (black).

the one step model, a photoelectron is produced and its kinetic energy E_{kin} is measured. By knowing the work function (φ), incident photon energy and kinetic energies of the photoelectrons, one may work backward to determine the binding energy dependence of the photoelectron intensity relative to the Fermi level (e.g., $E_{\text{Bind}} = h\nu - E_{\text{kin}} - \varphi$). The generic core-level photoemission intensity (from states with greater than approximately 20 eV binding energy) can be used to determine elemental composition, ion oxidation state(s), and/or coordination number(s) based on comparison to calibration data (see Fig. 9.4 left, as an example). The generic valence band photoemission intensity (spectral intensity between approximately 20 eV and the Fermi energy, E_{F}) can be used to make gross approximations of bonding character and the position of the valence band maximum relative to E_{F} (see Fig. 9.4 right). Overall, generic valence band photoemission is not useful in the study of magnetic exchange. However, by applying selection rule constraints to valance band photoemission through incident light polarization, angle of incidence, photon energy, angle of collection, and/or spin polarization of the collected photoelectrons, one can glean element, orbital, momentum, orientation and spin polarization information of the initial state that *is* very useful for understanding magnetic exchange and that is the subject of the following subsections.

9.4.1.1 Resonant photoemission

Although the core-level photoemission intensity (i.e., peak positions of spectra features) can be used to tell much about the oxidation state of ions in the MBM, it is those valence and conduction band states near the Fermi edge that ultimately provide the electronic structure information needed to understand magnetic exchange. One of the major hindrances of generic valence band photoemission is its inability to provide element specific information in addition to substantial energy overlap (hybridization) of the electronic bands/states; this deficiency can be partially overcome by tuning the incident photon energy so that there is a resonant enhancement of valence band states that are linked to their element specific core levels [100, 101]. The resonant photoemission (RPE) approach was first applied to the $M(TCNE)_x$ problem (for $M = V$) by Fahlman and coworkers in 2006 and the results are reproduced in Fig. 9.5 [83]. Prior to this study it was unknown whether the TCNE-based π^* or V 3d based t_{2g} dominated the highest occupied state and their degree of overlap/hybridization that would provide insight into the presumed strong superexchange. Further, the position of the π^* is important as it sets the energy of the lower Hubbard subband, from which the magnitude of the on-site Coulomb repulsion (from 0.5 to 2.0 eV) can be better estimated and validated [102, 103]. In the case of high-spin systems with more than three 3d electrons, the position of the t_{2g} and $e_g{}^*$ relative to the TCNE-based π and π^*, as determined by RPES [104], are even more important and the subject of future work.

X-ray emission spectroscopy (XES) offers a complementary means to extract RPE spectral intensity of the valence band. In the XES process, a valence electron fills a core hole, conserving energy by the emission of a photon—the energy of this exiting photon is sensitive to the local atomic environment as well as the valence electron distribution. However, a calibration photoemission spectrum must be collected in order to reference the XES results to

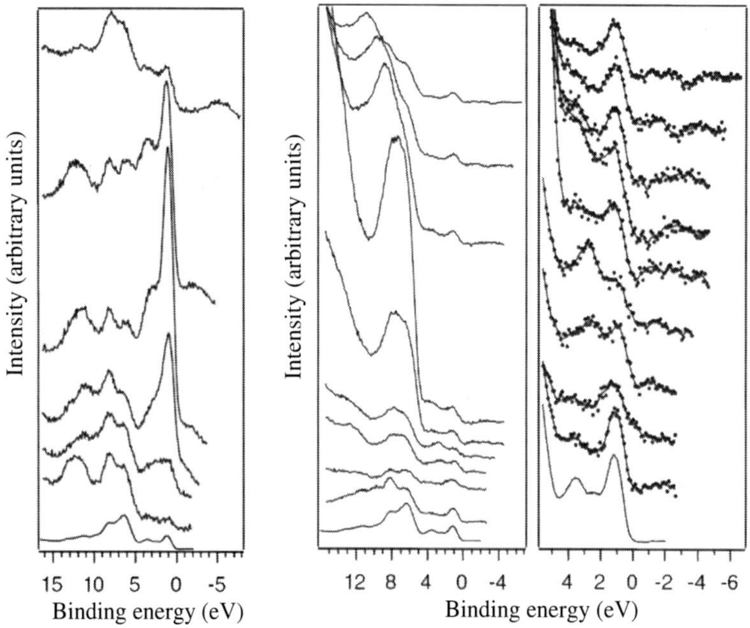

Figure 9.5 Resonant photoemission of a V(TCNE)$_x$ thin film over the (left) V *L3* and (right) C *K* edges. Adapted from Ref. [83].

E_F and the signal-to-noise ratio is often low. With next-generation synchrotron sources and elliptically polarized undulators, the ability to perform circularly polarized XES and/or circularly polarized resonant inelastic X-ray scattering (RIXS) may provide higher-quality and more exact overlap/hybridization information than RPE alone. Further, the scattering mechanism of RIXS is analogous and complementary to that of Raman spectroscopy (Section 9.5.3), whereby (in the case of the MTCNE) one can differentiate the π and π^* contribution to the exchange without the convolution of the central C=C of the TCNE.

9.4.1.2 Spin-resolved photoemission

Spin-polarized photoemission spectroscopy (SPPES) is used to resolve the spin-majority from spin-minority photoemission intensity; following kinetic energy analysis, the spin state of the photoelectron is determined [105]. SPPES of a pressed pellet of $[Fe^{II}(TCNE)(NCMe)_2][Fe^{III}Cl_4]$ was first reported (Fig 9.6 left) in 2009 by Caruso et al. [106] who determined that finite spin polarization and exchange splitting between the TCNE π^* and Fe $3d$ $e_g{}^*$ states exist within a few electron volts of E_F. More exact information regarding the hybridization and role of the t_{2g} with the TCNE π states could not be rigorously determined. Given the extreme surface sensitivity of low incident energy valence band photoemission, further advances in the use of SPPES of the $M[TCNE]_x$ complexes will require better in situ thin-film preparation methods, such as the ultrahigh vaccum (UHV)-compatible methods developed by Carlegrim [83, 107–109] and/or use of very-high-energy photoemission to achieve a greater effective probe depth. Second, given the poor transmission efficiency of a Mott polarimeter (10,000:1), a high fluence is required to collect quality spectra. However, one should remain wary of photoinduced changes that can occur to their material surface, especially molecule-based solids (see for example, Fig. 9.6 right).

9.4.1.3 Other photoemission studies

While measuring quantities proportional to the occupied DOS as a function of binding energy can reveal much about the local interactions of a solid, the picture is not complete without also mapping the unoccupied electronic structure. The unoccupied states, of interest to understanding exchange in MBMs, are allowed energy levels above the Fermi level between the conduction band edge and the vacuum level. In the case of the MTCNEs, the position, spin polarization and origin of the lowest unoccupied state(s) are important quantities as they help to corroborate/adjust the model(s) for the occupied states, their spin polarization, and assumed on-site Coulomb repulsion energy U. For example, does the unoccupied $e_g{}^*$ hybridize with the π^*+U and what is the role of electron correlation on the nearest neighbor? Spin-polarized inverse photoemission spectroscopy (SPIPES) is one means by which to extract these spin-dependent quantities In this method spin-polarized electrons are dropped into conduction band states as a function of incident kinetic energy while photons, released in

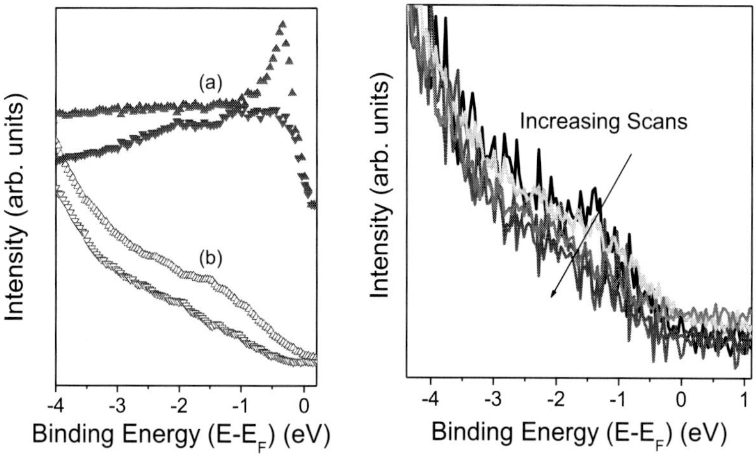

Figure 9.6 (Left) SPPES of clean cobalt (a) and $[Fe^{II}(TCNE)(NCMe)_2][Fe^{III}Cl_4]$ (b) demonstrating the resolution difference using the same spectrometer and incident photon energy. (Right) Photodegradation of a $[Fe^{II}(TCNE)(NCMe)_2][Fe^{III}Cl_4]$ pressed pellet as a function of photon fluence showing the importance of balancing the signal-to-noise ratio with the integral number of scans.

the energy conservation process, are measured to yield the inverse photoemission intensity [110]. However, the energy resolution and signal-to-noise ratio of SPIPES are poor and upkeep of the instrument is very difficult such that it should only be approached after exhausting all other techniques. A more promising technique is two-photon photoemission (2ppe) [111]; in this method, absorption of a *pump* photon causes an interband transition of a valence electron to a conduction band state. Prior to the de-excitation of the conduction band electron, a *probe* photon causes its ejection as a photoelectron. The kinetic energy of the photoelectron, less the energy of the probe photon, yields the binding energy of the unoccupied state relative to the vacuum level; if the work function is known, the binding energy may be determined relative to the Fermi level. Although the cross section for photoexcitation of the pumped electron is less than that of a ground-state electron, the technique can yield very high resolution; the major downfall in accessing spin information is the same as SPPES. If one can however quantum mechanically restrict the state to which the pumped electron resides by the absorption of a circularly polarized photon and utilize a spin-integrated electron energy analyzer for the probe photoelectrons, this may be the best of all techniques reported here. Until the spectroscopies as described above become more mature, a more accessible and prolific technique is that of circularly polarized X-ray absorption spectroscopy (XAS) (a.k.a., X-ray magnetic circular dichroism).

9.4.2 *X-Ray Absorption Spectroscopy*

In the X-ray absorption process a core-level electron is promoted to an unoccupied but bound state (below the vacuum level); the photon energy is swept while its transmission, promotion of electrons, or fluorescence yield is measured. In each case, information regarding the structure of the unoccupied DOS may be inferred on the basis of a well-understood core level. Like core-level photoemission, generic XAS can be used to extract quantitative oxidation and coordination information (including bond length), and it is element specific [112]. However, the limited selection rules afforded by linearly polarized light can make interpretation and inference of the near-Fermi-edge unoccupied electronic structure ambiguous (in the context of painting a magnetic exchange picture) as well as calibration to a binding energy scale. Again, through constraints such as light polarization and angle of incidence, it is found that spin resolution and quantities proportional to the partial density of unoccupied states can be more definitely assigned. In light of the above difficulties, full credit must be given to Fahlman and coworkers for their clever combination of generic XAS techniques with alkaline doping (Rb, Na) in the TCNE and M[TCNE]$_x$ systems to help extract where and to which orbitals the unpaired radical/polaronic weight exists (in energy) and/or to which it may be hybridized [103, 113].

9.4.2.1 X-ray magnetic circular dichroism

X-ray magnetic circular dichroism (XMCD) is an adaptation of XAS wherein absorption spectra are collected individually for left and right circularly polarized light (the light polarization may also be held constant while the magnetic field is switched). The difference in left and right absorption (or opposite external field application) is then plotted as a function of incident photon energy yielding a wealth of local and element-specific exchange information, some of which can be used to draw a picture of the unoccupied partial DOS (see Fig. 9.7). One of the advantages of XMCD over the photoemission techniques is the ability to apply an external magnetic field during the measurement (in the case of photon-in/photon-out analysis). This is a major help in extracting reasonable spin-based quantities given the signal-to-noise ratio from soft magnets such as the M[TCNE]$_x$ complexes. A disadvantage of XMCD is the inability to rigorously calibrate and place the measured states relative to the Fermi level; only in concert with inverse photoemission or 2ppe may one systematically rescale the photon energy to a binding energy axis. However, regardless of the energy axis used, the relative overlap of chemically distinct and dichroic states makes XMCD an important toolset for inferring exchange, bonding and related charge transfer information as first shown for the V[TCNE]$_x$ system by Kortright et al. [114] and de Jong et al. [107]. In this specific case however, the evidence from Kortright (Fig. 9.7) pointed toward a metal-to-ligand charge transfer (i.e., a π-interaction between electrons on the V3dt_{2g} manifold and π*-orbital of TCNE$^{\bullet-}$) as the basis for the strong antiferromagnetic exchange. Tengstedt and coworkers concluded oppositely that substantial ligand-to-metal charge transfer (resulting in a V^{2+} $3d$ ground state consists of a 60% $3d^3$ and 40% $3d^4L$ filling where L = hole on the TCNE$^{\bullet-}$ ligand) formed the exchange pathway foundation. To help rectify this disparity in seemingly valid interpretations of similar data, further toolsets that probed more definitely quantities were needed.

9.4.2.2 UV-visible magnetic circular dichroism

Ultraviolet-visible magnetic circular dichroism (UV/Vis MCD) is a commonly used technique in the biology community where chiral and transition moment containing molecules are studied [115]. There are few if any studies that apply this technique to the solid state. Like XMCD, UV/Vis MCD yields the differential

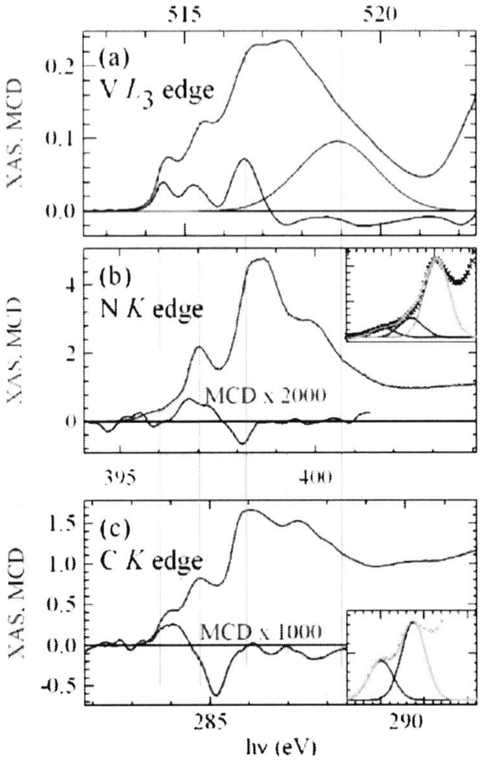

Figure 9.7 Vanadium (a), nitrogen (b) and carbon (c) absorption (red) and magnetic dichroism (blue) spectra from V[TCNE]$_x$, demonstrating hybrid states, and identification of the contributor to the lowest unoccupied state.

absorption of left and right circularly polarized light (or reversed magnetization directions with a single incident light polarization) but it is different in that it is not elementally distinct. The true power and differentiation of UV/Vis MCD from XMCD is that the spin-resolved final state can be mapped from a spin-polarized valence band state. By utilizing the $\Delta m_j = \pm 1$ selection rule, this means quantities proportional to the spin-resolved DOS may be extracted by comparing the measured dichroism spectra against either a known valence or conduction band. UV/Vis MCD results on the $Fe^{II}(TCNE^{\bullet-})(MeCN)_2\ SbF_6^-$ system are forthcoming.

9.4.3 *Raman and Infrared Absorption Spectroscopies*

The relationship between bonding and magnetic exchange in the $M[TCNE]_x$ systems remains elusive. Inferences of the mechanism(s) responsible for magnetic communication continue to be assigned without rigor by the spectroscopies described above. This is not a problem of the spectroscopies as much as it is for the nature of the samples under study. Raman spectroscopy however affords a "relatively" higher resolution probe of the magnetic communication pathway, despite its lack of circularly polarized light or spin-resolved electron analysis. In the case of the $M[TCNE]_x$ systems, by monitoring the Raman modes of the C=C and C≡N one can determine the amount of ligand-to-metal charge transfer, or vice versa, thus giving insight into the nature of the superexchange. By monitoring the frequencies of the characteristic vibration modes as a function of hydrostatic pressure the nature and degree of charge transfer can ultimately be related to the respective change in transition temperature and thus, the details of the magnetic exchange. Pokhodnya and coworkers have recently shown how this analysis has been used in the case of the $[M(TCNE)(NCMe)_2]X$ systems where M = Fe, Mn, or Ni and X = $FeCl_4$ or SbF_6 [40, 116].

9.5 Toward Predictive Understanding of Magnetic Exchange

The goal of this work is to call attention to outstanding challenges in the description and ultimate control over magnetic exchange in solids, with a focus on MBMs. The challenges encompass the phenomenological, computational and experimental fields that seek to relate the local electronic structure and space charge density with the trends observed in quantities such as transition temperature. While the phenomenological mean field analysis [117] provides for reasonable exchange constants for MBMs with different dimensionality of magnetic interaction, it does not account for or represent the means by which magnetic information may be communicated (or through which communication may be broken down) between states containing unpaired electrons

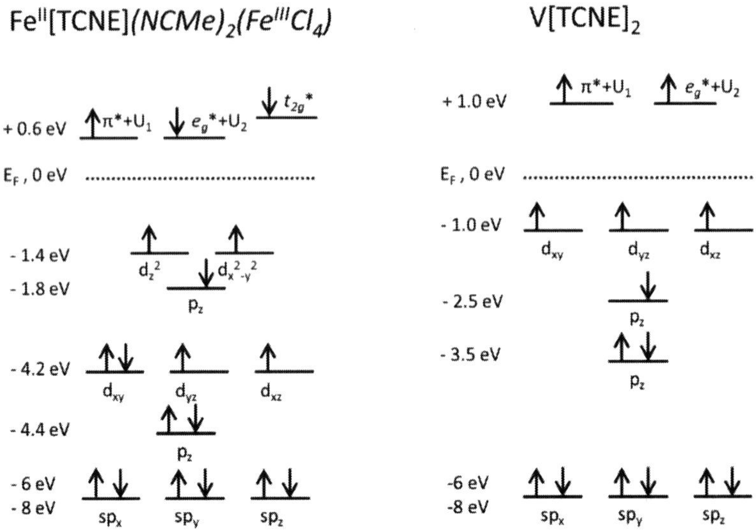

Figure 9.8 Energylevel and electronfilling scheme for the Fe[TCNE] $(NCMe)_2$ and $V[TCNE]_2$ complexes, based on experimentally derived evidence and phenomenological considerations of the strength of the on-site Coulomb repulsion.

(i.e., magnetically communicating bonds). In the same breath, computational/experimental tools that seek to calculate/measure the accurate magnitude and position of the spin-resolved partial DOS that allow for calculation of the exchange energy are sought The calculations definitely confirm the correlation between the interaction dimensionality and magnetic ordering temperatures, T_cs. However, the computational approach is not yet mature enough to provide reasonable quantitative correspondence between the calculated and experimentally observed T_cs. We leave this chapter with examples of possible orbital specific binding energy schemes for the $V[TCNE]_2$ and $Fe[TCNE](NCMe)_2$ complexes (Fig. 9.8) in an effort to emphasize the lack of understanding in describing magnetic communication pathways. The reader is reminded that the t_{2g} is classically considered to be non-bonding and the unpaired electronic weight for the TCNE is presented on the π*- or p_z-orbital in this case—through which pathways does the magnetic exchange occur and how may it be tuned?

References

1. C. Kittel, *Introduction to Solid State Physics*, 8th ed. Hoboken, New Jersey: John Wiley & Sons, 2004.
2. P. Fazekas, *Lecture Notes on Electron Correlation and Magnetism*. Singapore: World Scientific, 1999.
3. J. H. Van Vleck, *The Theory of Electric and Magnetic Susceptibilities*. Oxford: Oxford University Press, 1932.
4. I. Dzyaloshinsky, A thermodynamic theory of weak ferromagnetism of antiferromagnetics, *Phys. Chem. Solids* **4**(c), 241–255 (1958).
5. T. Moriya, New mechanism of anisotropic superexchange interaction, *Phys. Rev. Lett.*, **4**(5), 228–230 (1960).
6. D. Coffey, T. M. Rice, F. C. Zhang, Dzyaloshinskii-Moriya interaction in the cuprates, *Phys. Rev. B*, **44**(18), 10112–10116 (1991).
7. T. Moriya, Anisotropic superexchange interaction and weak ferromagnetism, *Phys. Rev.*, **249**(1949), 91–98 (1960).
8. W. Koshibae, Y. Ohta, S. Maekawa, Theory of Dzyaloshinski-Moriya antiferromagnetism in distorted CuO2 and NiO2 planes, *Phys. Rev. B*, **50**(6), 3767–3781 (1994).
9. O. Tchernyshyov, G. W. Chern, Spin-lattice coupling in frustrated antiferromagnets, *Physics*, **164**, 23 (2009).
10. C. Weber, F. Becca, F. Mila, Finite-temperature properties of frustrated classical spins coupled to the lattice, *Phys. Rev. B*, **72**, 6 (2005).
11. C. Jia, J. H. Han, Lattice-coupled antiferromagnet on frustrated lattices, *Phys. Rev. B*, **71**(21), 2–5 (2004).
12. W. Barford, R. J. Bursill, Peierls transition in the quantum spin-Peierls model, *Phys. Rev. Lett.*, **95**(13), 137207 (2005).
13. O. Kahn, *Molecular Magnetism*. Wiley-VCH Verlag GmbH, 1993.
14. R. Georges, J. J. Borras-Almenar, E. Coronado, J. Curely, M. Drillon, One-dimensional magnetism: an overview of the models, in *Magnetism: Molecules to Materials I: Models and Experiments*, Vol. 3, J. S. Miller, M. Drillon, Eds. Wiley-VCH Verlag GmbH & Co. KGaA, 2002.
15. K. Awaga, N. Wada, I. Watanabe, T. Inabe, Unusual crystal structures and properties of nitronylnitroxide radicals. Possible RVB states in molecule-based magnets, *Philos. Trans. R. Soc. A: Math., Phys. Eng. Sci.*, **357**(1762), 2893–2922 (1999).
16. R. L. Carlin, *Magnetochemistry*. Berlin, Heidelberg: Springer-Verlag, 1986.

17. J. S. Miller, A. J. Epstein, Organic and organometallic molecular magnetic materials: designer magnets, *Angew. Chem., Int. Ed. English*, **33**(4), 385–415 (1994).
18. D. A. Pejakovic, C. Kitamura, J. S. Miller, A. J. Epstein, Optical control of magnetic order in molecule-based magnet Mn(TCNE)x.y(CH2Cl2), *J. Appl. Phys.*, **91**, 7176–7178 (2002).
19. J.-W. Yoo, C.-Y. Chen, H. W. Jang, C. W. Bark, V. N. Prigodin, C. B. Eom, A. J. Epstein, Spin injection/detection using an organic-based magnetic semiconductor, *Nat. Mater.*, **9**(8), 638–642 (2010).
20. A. J. Epstein, Organic-based magnets: opportunities in photoinduced magnetism, spintronics, fractal magnetism, and beyond, *MRS Bull.*, **28**(7), 492–499 (2003).
21. J. Fabian, S. Das Sarma, Spintronics: fundamentals and applications, *Rev. Mod. Phys.*, **76**(2), 323–410 (2004).
22. A. R. Rocha, V. M. Garcia-Suarez, S. W. Bailey, C. J. Lambert, J. Ferrer, S. Sanvito, Towards molecular spintronics, *Nat. Mater.*, **4**, 335–339 (2005).
23. J. M. D. Coey, S. Sanvito, Magnetic semiconductors and half-metals, *J. Phys. D: Appl. Phys.*, **37**(7), 988–993 (2004).
24. P. a. Dowben, Are half-metallic ferromagnets half metals? (Invited), *J. Appl. Phys.*, **95**(11), 7453 (2004).
25. I. Ratera, J. Veciana, Playing with organic radicals as building blocks for functional molecular materials., *Chem. Soc. Rev.*, **41**(1), 303–349 (2012).
26. J. A. Crayston, J. N. Devine, J. C. Walton, Conceptual and synthetic strategies for the preparation of organic magnets, *Tetrahedron*, **56**(541), 7829–7857 (2000).
27. S. Lee, M. Yudkowsky, W. Halperin, M. Ogawa, B. Hoffman, One-dimensional magnetism in copper phthalocyanine, *Phys. Rev. B*, **35**(10), 5003–5007 (1987).
28. V. Y. Aristov, O. V. Molodtsova, V. V. Maslyuk, D. V. Vyalikh, V. M. Zhilin, Y. a Ossipyan, T. Bredow, I. Mertig, M. Knupfer, Electronic structure of the organic semiconductor copper phthalocyanine: experiment and theory, *J. Chem. Phys.*, **128**(3), 034703 (2008).
29. D. J. R. Brook, V. Lynch, B. Conklin, M. A. Fox, Spin delocalization in the copper(I) complexes of bis(verdazyl) diradicals, *J. Am. Chem. Soc.*, **119**(22), 5155–5162 (1997).
30. J. S. Miller, A. J. Epstein, Tetracyanoethylene-based organic magnets, *Chem. Commun.*, **1998**(13), 1319–1325 (1998).

31. J. S. Miller, A. J. Epstein, New higher-Tc molecular based magnets, *Synth. Met.*, **56**, 3291–3298 (1993).
32. J. S. Miller, A. J. Epstein, Organometallic electron-transfer salts with tetracyanoethylene exhibiting ferromagnetic coupling, *Adv. Chem. Ser.*, **226**, 419–432 (1990).
33. J. S. Miller, Magnetically ordered molecule-based materials, *Chem. Soc. Rev*, **40**(6), 3266–3296 (2011).
34. E. Ruiz, A. Rodríguez-Fortea, S. Alvarez, M. Verdaguer, Is it possible to get high T(C) magnets with Prussian blue analogues? A theoretical prospect, *Chem. (Weinheim Bergstr., Germany)*, **11**(7), 2135–2144 (2005).
35. H. Tokoro, S. Ohkoshi, Novel magnetic functionalities of Prussian blue analogs, *Dalton Trans.*, **40**(26), 6825–6833 (2011).
36. M. Verdaguer, G. Girolami, Magnetic prussian blue analogs, in *Magnetism: Molecules to Materials V*, 283–346, J. S. Miller, M. Drillon, Eds. Weinheim: Wiley-VCH Verlag GmbH & Co. KGaA, 2005.
37. J. W. Raebiger, J. S. Miller, Magnetic ordering in the rare earth molecule-based magnets, Ln(TCNE)(3) (Ln = Gd, Dy; TCNE = tetracyanoethylene), *Inorg. Chem.*, **41**(12), 3308–3312 (2002).
38. X.-Y. Wang, C. Avendaño, K. R. Dunbar, Molecular magnetic materials based on 4d and 5d transition metals, *Chem. Soc. Rev*, **40**(6), 3213–3238 (2011).
39. J. S. Miller, J. C. Calabrese, R. S. McLean, A. J. Epstein, meso-(Tetraphenylporphinato)manganese(III)-tetracyanoethenide, [MnIIITPP]::⊕[TCNE].⊖. A new structure-type linear-chain magnet with a Tc of 18K, *Adv. Mater.*, **4**(7–8), 498–501 (1992).
40. C. Olson, C. L. Heth, S. H. Lapidus, P. W. Stephens, G. J. Halder, K. Pokhodnya, Spectroscopic study of (two-dimensional) molecule-based magnets: [M(II)(TCNE)(NCMe)2][SbF6] (M = Fe, Mn, Ni), *J. Chem. Phys.*, **135**(2), 024503 (2011).
41. K. H. Stone, P. W. Stephens, A. C. McConnell, E. Shurdha, K. I. Pokhodnya, J. S. Miller, Mn(II)(TCNE)(3/2)(I3)(1/2)–a 3D network-structured organic-based magnet and comparison to a 2D analog, *Adv. Mater. Deerfield Beach Fla.*, **22**(23), 2514–2519 (2010).
42. R. Sessoli, D. Gatteschi, A. Caneschi, M. Novak, Magnetic bistability in a metal-ion cluster, *Nature*, **365**, 141–143 (1993).
43. L. Thomas, F. Lionti, R. Ballou, D. Gatteschi, Macroscopic quantum tunnelling of magnetization in a single crystal of nanomagnets, *Nature*, **383**, 145–147 (1996).

44. E. K. Brechin, A. Graham, P. E. Y. Milne, M. Murrie, S. Parsons, R. E. P. Winpenny, New high-spin clusters featuring transition metals, *Phil. Trans. R. Soc. Lond. A*, **357**, 3119–3137 (1999).

45. F. Neese, D. A. Pantazis, What is not required to make a single molecule magnet, *Faraday Discuss.*, **148**, 229–238; discussion 299–314 (2011).

46. D. Gatteschi, Single molecule magnets: a new class of magnetic materials, *J. Alloys Compd.*, **317**–318, 8–12 (2001).

47. J. D. Rinehart, J. R. Long, Exploiting single-ion anisotropy in the design of f-element single-molecule magnets, *Chem. Sci.*, **2**(11), 2078–2085 (2011).

48. H. Kuhn, A quantum-mechanical theory of light absorption of organic dyes and similar compounds, *J. Chem. Phys.*, **17**(12), 1198 (1949).

49. N. F. Mott, The basis of the electron theory of metals, with special reference to the transition metals, *Proc. Phys. Soc., Sect. A*, **62**(7), 416–422 (1949).

50. J. Hubbard, Electron correlations in narrow energy bands, *Proc. R. Soc. A: Math., Phys. Eng. Sci.*, **276**(1365), 238–257 (1963).

51. J. Kortright, D. Lincoln, R. Edelstein, A. J. Epstein, Bonding, backbonding, and spin-polarized molecular orbitals: basis for magnetism and semiconducting transport in V[TCNE]x∼2, *Phys. Rev. Lett.*, **100**(25), 2–5 (2008).

52. J. P. Perdew, Jacob's ladder of density functional approximations for the exchange-correlation energy, *AIP Conf. Proc.*, **577**(1), 1–20 (2001).

53. J. Pople, Nobel lecture: quantum chemical models, *Rev. Mod. Phys.*, **71**(5), 1267–1274 (1999).

54. J. Kohanoff, *Electronic Structure Calculations for Solids and Molecules*. Cambridge: Cambridge University Press, 2006.

55. R. M. Martin, *Electronic Structure: Basic Theory and Practical Methods*. Cambridge University Press, 2004.

56. A. Szabo, N. S. Ostlund, *Modern Quantum Chemistry*. Mineola, NY: Dover, 1996.

57. P. Fulde, *Correlated Electrons in Quantum Matter*. Hackensack: World Scientific, 2012.

58. J. P. Perdew, A. Ruzsinszky, J. Tao, V. N. Staroverov, G. E. Scuseria, G. I. Csonka, Prescription for the design and selection of density functional approximations: more constraint satisfaction with fewer fits, *J. Chem. Phys.*, **123**(6), 062201–062201-9 (2005).

59. S. Kümmel, L. Kronik, Orbital-dependent density functionals: theory and applications, *Rev. Mod. Phys.*, **80**(1), 3–60 (2008).
60. Electronic correlation, *Wikipedia, the free encyclopedia*. Apr. 2013.
61. Exchange interaction, *Wikipedia, the free encyclopedia*. May 2013.
62. R. Nesbet, Approximate methods in the quantum theory of many-Fermion systems, *Rev. Mod. Phys.*, **33**(1), 28–36 (1961).
63. G. H. Booth, A. Grüneis, G. Kresse, A. Alavi, Towards an exact description of electronic wavefunctions in real solids, *Nature*, **493**(7432), 365–370 (2013).
64. D. M. Ceperley, Ground state of the electron gas by a stochastic method, *Phys. Rev. Lett.*, **45**(7), 566–569 (1980).
65. E. Wigner, On the interaction of electrons in metals, *Phys. Revi*, **46**(11), 1002–1011 (1934).
66. L. Hedin, B. I. Lundqvist, Explicit local exchange-correlation potentials, *J. Phys. C: Solid State Phys.*, **4**(14), 2064–2083 (1971).
67. U. von Barth, L. Hedin, A local exchange-correlation potential for the spin polarized case. i, *J. Phys. C: Solid State Phys.*, **5**(13), 1629–1642 (1972).
68. O. Gunnarsson, B. Lundqvist, Exchange and correlation in atoms, molecules, and solids by the spin-density-functional formalism, *Phys. Rev. B*, **13**(10), 4274–4298 (1976).
69. J. P. Perdew, Self-interaction correction to density-functional approximations for many-electron systems, *Phys. Rev. B*, **23**(10), 5048–5079 (1981).
70. A. D. Becke, On the large-gradient behavior of the density functional exchange energy, *J. Chem. Phys.*, **85**(12), 7184 (1986).
71. A. D. Becke, Correlation energy of an inhomogeneous electron gas: a coordinate-space model, *J. Chem. Phys.*, **88**(2), 1053 (1988).
72. J. P. Perdew, Y. Wang, Accurate and simple analytic representation of the electron-gas correlation energy, *Phys. Rev. B*, **45**(23), 13244–13249 (1992).
73. A. D. Becke, A new mixing of Hartree–Fock and local density-functional theories, *J. Chem. Phys.*, **98**(2), 1372 (1993).
74. J. P. Perdew, K. Burke, Comparison shopping for a gradient-corrected density functional, *Int. J. Quantum Chem.*, **57**(3), 309–319 (1996).
75. A. D. Becke, Density functionals for static, dynamical, and strong correlation, *J. Chem. Phys.*, **138**(7), 074109 (2013).

76. K. Burke, Perspective on density functional theory, *J. Chem. Phys.*, **136**(15), 150901 (2012).
77. J. P. Perdew, M. Ernzerhof, K. Burke, Rationale for mixing exact exchange with density functional approximations, *J. Chem. Phys.*, **105**(22), 9982–9985 (1996).
78. M. Jain, J. R. Chelikowsky, S. G. Louie, Reliability of hybrid functionals in predicting band gaps, *Phys. Rev. Lett.*, **107**(21), 216806 (2011).
79. D. I. Bilc, R. Orlando, R. Shaltaf, G.-M. Rignanese, J. Íñiguez, P. Ghosez, Hybrid exchange-correlation functional for accurate prediction of the electronic and structural properties of ferroelectric oxides, *Phys. Rev. B*, **77**(16), 165107 (2008).
80. J. Graciani, A. M. Marquez, J. J. Plata, Y. Ortega, N. C. Hernandez, A. Meyer, C. M. Zicovich-Wilson, J. F. Sanz, Comparative study on the performance of hybrid DFT functionals in highly correlated oxides: the case of CeO2 and Ce2O3, *J. Chem. Theory Comput.*, **7**(1), 56–65 (2011).
81. R. Peverati, D. G. Truhlar, The quest for a universal density functional: The accuracy of density functionals across a broad spectrum of databases in chemistry and physics, *Philos. Trans. R. Soc. A*, **372**, 20120476/1-51 (2014).
82. S. Erdin, Ab initio studies of tetracyanoethylene-based organic magnets, *Phys. B: Condens. Matter*, **403**(10)–11, 1964–1970 (2008).
83. C. Tengstedt, M. P. De Jong, A. Kanciurzewska, E. Carlegrim, M. Fahlman, X-ray magnetic circular dichroism and resonant photomission of V(TCNE)x hybrid magnets, *Phys. Rev. Lett.*, **96**(5), 057209 (2006).
84. V. I. Anisimov, J. Zaanen, O. K. Andersen, Band theory and Mott insulators: Hubbard U instead of Stoner I, *Phys. Rev. B*, **44**(3), 943–954 (1991).
85. R. Hafner, Ab-initio simulations of materials using VASP: density-functional theory and beyond, *J. Comput. Chem.*, **29**(13), 2044–2078 (2008).
86. S. Yamanaka, K. Yamaguchi, Ab initio extended density functional theory for strongly correlated electron systems: fundamental aspects of the broken-symmetry approach and possible applications for molecular material design, *Bull. Chem. Soc. Jpn.*, **77**(7), 1269–1286 (2004).
87. P. E. Blochel, Projector augmented-wave method, *Phys. Rev. B*, **50**(24), 17953–17959 (1994).

88. M. Cococcioni, S. de Gironcoli, Linear response approach to the calculation of the effective interaction parameters in the LDA+U method, *Phys. Rev. B*, **71**(3), 035105 (2005).

89. N. J. Mosey, P. Liao, E. a Carter, Rotationally invariant ab initio evaluation of Coulomb and exchange parameters for DFT+U calculations, *J. Chem. Phys.*, **129**(1), 014103 (2008).

90. M. S. Hybertsen, M. Schlüter, N. E. Christensen, Calculation of Coulomb-interaction parameters for La2CuO4 using a constrained-density-functional approach, *Phys. Rev. B*, **39**(13), 9028–9041 (1989).

91. W. Shum, A. Epstein, J. Miller, Spin-polarized electronic structure for the layered two-dimensional [FeII(TCNE)(NCMe)2][FeIIICl4] organic-based magnet, *Phys. Rev. B*, **80**(6), 1–6 (2009).

92. J. C. Wojdeł, I. D. P. R. Moreira, S. T. Bromley, F. Illas, Prediction of half-metallic conductivity in Prussian blue derivatives, *J. Mater. Chem.*, **19**(14), 2032 (2009).

93. P. M. Panchmatia, B. Sanyal, P. M. Oppeneer, GGA+U modeling of structural, electronic, and magnetic properties of iron porphyrin-type molecules, *Chem. Phys.*, **343**(1), 47–60 (2008).

94. A. P. Weber, A. N. Caruso, E. Vescovo, M. E. Ali, K. Tarafder, S. Z. Janjua, J. T. Sadowski, P. M. Oppeneer, Magnetic coupling of Fe-porphyrin molecules adsorbed on clean and c(2 × 2) oxygen-reconstructed Co(100) investigated by spin-polarized photoemission spectroscopy, *Phys. Rev. B*, **87**(18), 184411 (2013).

95. P. M. Panchmatia, M. E. Ali, B. Sanyal, P. M. Oppeneer, Halide ligated iron porphines: a DFT+U and UB3LYP study, *J. Phys. Chem. A*, **114**(51), 13381–13387 (2010).

96. I. Ciofini, DFT calculations of molecular magnetic properties of coordination compounds, *Coord. Chem. Rev.*, **238**–239, 187–209 (2003).

97. J. J. Novoa, M. Deumal, J. Jornet-Somoza, Calculation of microscopic exchange interactions and modelling of macroscopic magnetic properties in molecule-based magnets, *Chem. Soc. Rev*, **40**, 3182–3212 (2011).

98. E. Ressouche, J. Schweizer, Ab initio calculations versus polarized neutron diffraction for the spin density of free radicals, *Monatsh. Chem.*, **134**(2), 235–253 (2003).

99. B. K. I. Pokhodnya, A. J. Epstein, J. S. Miller, Thin-film V[TCNE]x magnets, *Adv. Mater.*, **12**(6), 410–413 (2000).

100. F. de Groot, X-ray absorption of transition metal oxides: an overview of the theoretical approaches, *J. Electron Spectrosc. Relat. Phenom.*, **62**, 111–130 (1993).

101. F. De Groot, Multiplet effects in X-ray spectroscopy, *Coord. Chem. Rev.*, **249**(1)–2, 31–63 (2005).
102. C. Tengstedt, M. Unge, M. de Jong, S. Stafström, W. Salaneck, M. Fahlman, Coulomb interactions in rubidium-doped tetracyanoethylene: a model system for organometallic magnets, *Phys. Rev. B*, **69**(16), 1–7 (2004).
103. E. Carlegrim, B. Gao, A. Kanciurzewska, M. de Jong, Z. Wu, Y. Luo, M. Fahlman, Near-edge x-ray absorption studies of Na-doped tetracyanoethylene films: a model system for the V(TCNE)x room-temperature molecular magnet, *Phys. Rev. B*, **77**(5), 1–8 (2008).
104. E. Carlegrim, Y. Zhan, M. P. de Jong, M. Fahlman, Electronic structure of thin film cobalt tetracyanoethylene, Co(TCNE)x, *Synth. Met.*, **161**(17)–18, 1892–1897 (2011).
105. P. D. Johnson, Spin-polarized photoemission, *Rep. Prog. Phys.*, **60**(11), 1217–1304 (1997).
106. A. Caruso, K. Pokhodnya, W. Shum, W. Ching, B. Anderson, M. Bremer, E. Vescovo, P. Rulis, a. Epstein, J. Miller, Direct evidence of electron spin polarization from an organic-based magnet: [FeII(TCNE)(NCMe)2][FeIIICl4], *Phys. Rev. B*, **79**(19), 1–5 (2009).
107. M. de Jong, C. Tengstedt, A. Kanciurzewska, E. Carlegrim, W. Salaneck, M. Fahlman, Chemical bonding in V(TCNE)x (x$\sim$2) thin-film magnets grown in situ, *Phys. Rev. B*, **75**(6), 1–7 (2007).
108. P. Bhatt, A. Kanciurzewska, E. Carlegrim, M. Kapilashrami, L. Belova, K. V Rao, M. Fahlman, Ferromagnetism above room temperature in nickel–tetracyanoethylene thin films, *J. Mater. Chem.*, **19**(36), 6610 (2009).
109. E. Carlegrim, *Development of Organic-based Thin Film Magnets for Spintronics*, Linköping University, 2010.
110. J. Unguris, A. Seiler, R. Celotta, D. Pierce, P. Johnson, N. Smith, Spin-polarized inverse photoelectron spectroscopy of solid surfaces: Ni(110), *Phys. Rev. Lett.*, **49**(14), 1047–1050 (1982).
111. H. Ueba, B. Gumhalter, Theory of two-photon photoemission spectroscopy of surfaces, *Prog. Surf. Sci.*, **82**(4)–6, 193–223 (2007).
112. D. Haskel, Z. Islam, J. Lang, C. Kmety, G. Srajer, K. Pokhodnya, A. Epstein, J. Miller, Local structural order in the disordered vanadium tetracyanoethylene room-temperature molecule-based magnet, *Phys. Rev. B*, **70**(5), 1–9 (2004).
113. E. Carlegrim, A. Kanciurzewska, M. P. de Jong, C. Tengstedt, M. Fahlman, The unoccupied electronic structure of the semi-conducting room

temperature molecular magnet V(TCNE)2, *Chem. Phys. Lett.*, **452**(1)-3, 173–177 (2008).

114. J. Kortright, D. Lincoln, R. Edelstein, and a. Epstein, Bonding, Backbonding, and Spin-Polarized Molecular Orbitals: Basis for Magnetism and Semiconducting Transport in V[TCNE]x~2, *Phys. Rev. Lett.*, **100**(25), 2–5 (2008).
115. J. C. Sutherland, Measuurement of circular dichroism and related spectroscopies with conventional and synchrotron light sources: theory and instrumentation, in *Modern Techniques for Circular Dichroism Spectroscopy*, B. A. Wallace, R. W. Janes, Eds. Amsterdam: IOS Press, 2009.
116. A. E. Midgley, C. Olson, C. L. Heth, A. N. Caruso, M. B. Kruger, G. J. Halder, J. A. Schlueter, K. Pokhodnya, Chemical bonding and magnetic exchange in two-dimensional [M(TCNE)(NCMe)(2)]X (M = Fe, Mn; X = FeCl4, SbF6) magnets: a pressure study, *J. Chem. Phys.*, **138**(1) (2013).
117. A. C. Mcconnell, R. S. Fishman, J. S. Miller, Mean field analysis of the exchange coupling (J) for two- and three-dimensional structured tetracyanoethenide ($TCNE^{•-}$)-based magnets, *J. Phys. Chem. C*, **116**, 16154–16160 (2012).

Chapter 10

Spin-Polarized Scanning Tunneling Microscopy

Kangkang Wang[a,*] and Saw-Wai Hla[a,b]

[a]*Center for Nanoscale Materials, Argonne National Laboratory, 9700 S Cass Avenue, Lemont, IL 60439, USA*

[b]*Nanoscale and Quantum Phenomena Institute, Physics and Astronomy Department, Ohio University, Athens, OH 45701, USA*

hla@ohio.edu

As material sizes shrink, novel magnetic phenomena often emerge that are drastically different from their bulk counterparts. Spin-polarized scanning tunneling microscopy enables studying of magnetic phenomena on surfaces with real space imaging and spectroscopy techniques down to the atomic level. Because of its ability to directly correlate topographic, electronic, and magnetic properties, it is a powerful tool to resolve and even manipulate complex spin structures at the ultimate spatial resolution.

---

*Currently at the Magnetic Metrology and Media Physics (M3P) group, Recording Media Operations, at Seagate Technologies, Fremont, California

---

Handbook of Nanomagnetism: Applications and Tools
Edited by Rosa A. Lukaszew

ISBN 978-981-4613-04-0 (Hardcover), 978-981-4613-05-7 (eBook)
www.panstanford.com

10.1 Introduction

In a conventional scanning tunneling microscope (STM) [1], an extremely sharp conducting tip is placed at the proximity of the material surface, less than 1 nm away. When a voltage is applied across the tip–surface junction, electrons can tunnel between the surface and the tip. The tunneling current is used to control the tip–sample distance as well as to probe the local density of states (LDOS) of the sample surface. In the case of a spin-polarized (SP)-STM, a *magnetized* tip is used to probe the magnetic structure of the sample surface [2, 3]. The operation of the SP-STM is somewhat analogous to tunneling magnetoresistance (TMR) [4]. In TMR, an ultrathin insulating film is sandwiched between two ferromagnetic materials and the tunneling current intensity between them is dependent on the parallel (increased intensity) or antiparallel (decreased intensity) spin configurations of the two magnetic electrodes. In an SP-STM, the tip and the sample act as two magnetic electrodes, while the space between them is the tunneling barrier (Fig. 10.1). The magnitude of the SP tunneling current depends on the relative

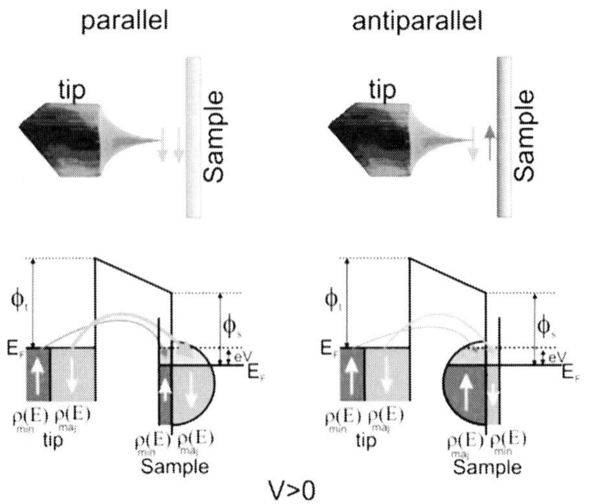

Figure 10.1 Schematic drawings showing electron tunneling between a magnetic tip and a magnetic sample for parallel and antiparallel tip and sample magnetizations (left and right, respectively).

magnetization directions between the tip and the sample. The LDOS of the sample (or the tip) at any voltage have spin polarization that can be defined as

$$\rho\,(\mathrm{E}) \equiv \frac{\rho_{\mathrm{maj}}\,(\mathrm{E}) - \rho_{\mathrm{min}}(\mathrm{E})}{\rho_{\mathrm{maj}}\,(\mathrm{E}) + \rho_{\mathrm{min}}(\mathrm{E})} \tag{10.1}$$

where $\rho_{\mathrm{maj}}(\mathrm{E})$ and $\rho_{\mathrm{min}}(\mathrm{E})$ denote the DOSs of the majority and minority electrons. In ferromagnetic materials for example, the band is split into two subbands called minority and majority bands due to exchange interactions. The spin polarization defined above is then a function of energy and has a maximum value of 1 for 100% spin polarization. The energy-dependent spin-summed DOS can be expressed as

$$\rho\,(\mathrm{E}) = \rho_{\mathrm{maj}}\,(\mathrm{E}) + \rho_{\mathrm{min}}(\mathrm{E}) \tag{10.2}$$

A common approach to describe SP tunneling is an extension of the Tersoff–Hamann [5] model to include the spin-dependent tunneling part. Here, it is assumed that the spin of an electron is conserved during tunneling. The sample DOS is also split into majority and minority bands, which are shifted by a finite amount in energy. The SP tunneling current for the parallel spin case can be expressed as

$$I_{\mathrm{sp}}^{\uparrow\uparrow} \propto \rho_{\mathrm{t}}^{\mathrm{maj}} \cdot \tilde{\rho}_{\mathrm{s}}^{\mathrm{maj}}\,(eV) + \rho_{\mathrm{t}}^{\mathrm{min}} \cdot \tilde{\rho}_{\mathrm{s}}^{\mathrm{min}}\,(eV) \tag{10.3}$$

And for the antiparallel case

$$I_{\mathrm{sp}}^{\uparrow\downarrow} \propto \rho_{\mathrm{t}}^{\mathrm{maj}} \cdot \tilde{\rho}_{\mathrm{s}}^{\mathrm{min}}\,(eV) + \rho_{\mathrm{t}}^{\mathrm{min}} \cdot \tilde{\rho}_{\mathrm{s}}^{\mathrm{maj}}\,(eV) \tag{10.4}$$

where $\rho_{\mathrm{t}}^{\mathrm{maj}}$ and $\rho_{\mathrm{t}}^{\mathrm{min}}$ are the SP DOSs of the tip, and the SP DOSs of the sample integrated over an energy window are

$$\tilde{\rho}_{\mathrm{s}}^{\mathrm{maj}}\,(eV) = \int_0^{eV} \rho_{\mathrm{s}}^{\mathrm{maj}}(E_{\mathrm{F}} + \epsilon)T\,(d,\, E_{\mathrm{F}} + \epsilon,\, V)d\epsilon \tag{10.5}$$

$$\tilde{\rho}_{\mathrm{s}}^{\mathrm{min}}\,(eV) = \int_0^{eV} \rho_{\mathrm{s}}^{\mathrm{min}}(E_{\mathrm{F}} + \epsilon)T\,(d,\, E_{\mathrm{F}} + \epsilon,\, V)d\epsilon \tag{10.6}$$

In these equations, T is the transmission coefficient, d is the distance between the sample and the tip, E_{F} is the Fermi energy, and V is the applied bias.

One can combine both cases and write in a more general term as

$$I_{\text{sp}}(\text{E}) \propto \frac{I_{sp}^{\uparrow\uparrow} + I_{\text{sp}}^{\uparrow\downarrow}}{2} \pm \frac{I_{sp}^{\uparrow\uparrow} - I_{\text{sp}}^{\uparrow\downarrow}}{2} \tag{10.7}$$

By introducing the energy-dependent magnetization

$$\mathbf{m}_{\text{t}}(\text{E}) = \rho_{\text{t}}^{\text{maj}}(\text{E}) - \rho_{\text{t}}^{\text{min}}(\text{E}) \tag{10.8}$$

$$\mathbf{m}_{\text{s}}(\text{E}) = \rho_{\text{s}}^{\text{maj}}(\text{E}) - \rho_{\text{s}}^{\text{min}}(\text{E}) \tag{10.9}$$

$$\begin{aligned}\tilde{\boldsymbol{m}}_{\text{s}}(eV) &= \int_0^{eV} \mathbf{m}_{\text{s}}(E_{\text{F}} + \epsilon) \cdot T(d, E_{\text{F}} + \epsilon, eV)\, d\epsilon \\ &= \tilde{\rho}_{\text{s}}^{\text{maj}}(eV) - \tilde{\rho}_{\text{s}}^{\text{min}}(eV)\end{aligned} \tag{10.10}$$

Substituting these new energy-dependent magnetization terms, the spin-dependent tunneling current can then be expressed as

$$I_{\text{sp}}(V) \propto \rho_{\text{t}} \cdot \tilde{\rho}_s(eV) \pm \mathbf{m}_{\text{t}} \cdot \tilde{\boldsymbol{m}}_{\text{s}}(eV) \tag{10.11}$$

The plus or minus sign depends on whether the two magnetization directions are parallel or antiparallel. For arbitrary relative orientations between the two magnetization directions, one can easily generalize the derivation above and obtain

$$I_{\text{sp}}(V) \propto \rho_{\text{t}} \cdot \tilde{\rho}_s(eV) \pm \mathbf{m}_{\text{t}} \cdot \tilde{\boldsymbol{m}}_{\text{s}}(eV) \cdot \cos\theta \tag{10.12}$$

where θ is the angle between the two magnetization directions. The first term corresponds to a spin-averaged part, while the second term corresponds to an SP part. One can also propagate a similar analysis for the differential conductance and obtain the SP spectroscopy (dI/dV) expression

$$\frac{dI}{dV}_{\text{sp}}(V) \propto \rho_{\text{t}} \cdot \rho_{\text{s}}(E_{\text{F}} + eV) \pm \mathbf{m}_{\text{t}} \cdot \mathbf{m}_{\text{s}}(E_{\text{F}} + eV) \cdot \cos\theta \tag{10.13}$$

Equations 10.13 and 10.14 are the key for the SP tunneling current imaging and SP dI/dV mapping, respectively. The SP parts of both equations include a cos θ term. This means that a maximum SP contrast can be achieved when $\theta = 0°$ or $180°$, which represents parallel and antiparallel magnetizations between the tip and sample magnetization directions. When the tip and sample magnetization directions are perpendicular to each other, that is, $\theta = 90°$, the spin contrast vanishes. Thus, the SP current depends on the relative orientation between the tip and the sample magnetization. In the case of a parallel spin configuration between the tip and the sample, both minority and majority tunneling channels generate a larger tunneling current, while the tunneling current intensity reduces in the antiparallel spin configuration due to fewer available spin states on the other electrode. This provides an scanning tunneling microscopy image contrast as brighter and darker for parallel and antiparallel tip–sample spin configurations, respectively.

In addition to the SP component, the tunneling current also contains a spin-averaged component. Therefore, the spin-dependent information is intermingled with the spin-averaged topographic information, and sometimes spin contrast can be hard to achieve. In such cases, the SP conductance map can be utilized because it is not dependent on the topography but sensitive only to the LDOS. In some other cases applying an external magnetic field can aid in clarifying the origin of the spin state. The magnetic field could in principle modify the tip or the sample magnetization, as well causing concomitant change in the magnetic contrast. SP scanning tunneling microscopy enables us not only to image the parallel and antiparallel spin contrasts but also to directly unveil novel spin configurations, such as skyrmions [6, 7], spin spirals [8–11], and 3D spin pyramids [12]. In addition to studying magnetic ground states that are relatively stable, spin-polarized tunneling spectroscopy (SP-STS) has been used to probe magnetization dynamics down to the atomic scale [14].

10.2 Preparation of the SP-STM Tip

The first and foremost requirement for SP imaging is to have an SP-STM tip. Conventionally, tungsten (W) and platinum–iridium (Pt/Ir) tips are the two most widely used probes for scanning tunneling microscopy purposes. However, none of these materials is magnetic. In this section, we will briefly review several methods to prepare SP probes for SP scanning tunneling microscopy.

The most straightforward way to prepare an SP-STM tip is to make tips from bulk magnetic materials such as iron (Fe) and nickel (Ni). Polycrystalline wires of these magnetic materials are first electrochemically etched to form a thin neck and then pulled apart to form a clean tip apex. Alternatively, a tip apex could be obtained by electrochemical etching and then cleaned further by ion sputtering or electron-beam heating in an ultrahigh vacuum (UHV) environment. An advantage of this approach is that the spin polarization of the tip material is known, and the tip will remain magnetic even after it loses materials via modifications during tunneling. On the other hand, a main disadvantage of using a bulk magnetic tip is its high stray magnetic field, which can easily perturb the magnetic structures of the sample surface. In addition, since the atomically precise shape of the tip apex is unknown, the magnetization direction of the tip is not well defined.

To reduce the stray field from the tip, bulk antiferromagnetic materials such as chromium (Cr) have been used to make SP-STM tips [15, 16]. Because of the antiferromagnetic nature of Cr, the stray field at the sample is greatly reduced compared to the bulk ferromagnetic tips. However, an exact magnetization direction of these bulk antiferromagnetic tips is usually not known and calibration procedures are often required for SP scanning tunneling microscopy experiments. Generally, the magnetization direction of the tip can be determined by using known test sample having both *in-plane* and *out-of-plane* magnetization components.

A more common approach to make SP-STM tips is to coat a nonmagnetic tip material such as W or Pt/Ir with a thin layer of magnetic materials. A general procedure for preparing such tips is as following (Fig. 10.2); a polycrystalline W wire is electrochemically etched to form a sharp tip, and then it is flash-annealed in UHV up to 2000°C to remove surface oxides as well as to slightly melt the tip apex. Then a certain thickness of a magnetic material (e.g., Fe, Cr) is coated onto the tip apex. Because the tip gets slightly blunt (tip radius $>$200 nm), the magnetization direction of the thin coated layer (generally $<$10 nm) is mostly governed by the material specific film anisotropies rather than the shape anisotropy as in the case of bulk tips. For example, $>$3 mL of Fe-coated W tips generally exhibit an in-plane magnetization direction, while 5–10 mL of gadolinium (Gd)-coated W tips have an out-of-plane magnetization direction [17]. An out-of-plane magnetization is also reported for the case of a thin cobalt (Co) film coated on a gold cover W tip [18]. While the magnetic coating is generally very thin,

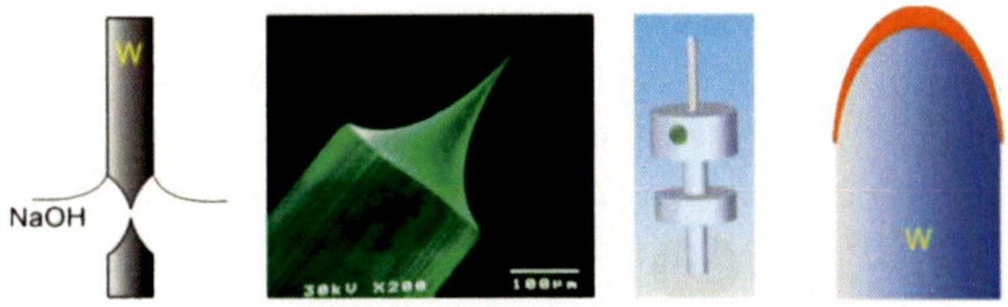

Figure 10.2 Preparation of coated magnetic tips. From left to right: Electrochemical etching of a W tip, a scanning electron microscopy image of the tip apex, a tip mounted to a tip holder, and deposition of magnetic material (red) onto the tip apex.

it still has a finite stray field that is large enough to perturb the magnetism of the sample surface, such as the domain wall positions. Antiferromagnetic materials like Cr can be used for the magnetic coating of the tip to reduce unwanted magnetic interactions [19]. An in situ transmission electron microscope investigation shows that Cr can grow epitaxially on a W tip but the magnetization of the nanocrystalline Cr at the W tip apex is difficult to control [20]. The ferromagnetic tips, on the other hand, have their own advantage as compared to antiferromagnetic tips. The main advantage is that an external magnetic field (from a superconducting magnet or a permanent magnet) can be applied to the tip–sample junction in order to align the tip's magnetization direction. For this purpose the field strength is usually selected to be large enough to modify the tip's magnetization but small enough to not perturb the sample's magnetic structure.

Another way to change from nonmagnetic into an SP-STM tip is by simply dipping the tip into a softer magnetic material. The tip then picks up magnetic materials from the surface and it can become a magnetic tip [21]. A voltage pulse applied between the tip and the sample can also transfer some of the magnetic materials to the tip apex, thereby resulting to form a magnetic tip [22]. In addition, atom manipulation techniques can also be used to pick up magnetic atoms from the surface to the tip apex to form a magnetic tip as well [23].

Another unique type of SP-STM probe has a tiny coil wrapped around either a soft magnetic tip or a ring-shaped probe [24–27]. When an alternating current is applied through the coil, an alternating magnetization at the tip apex can be induced. The frequency of this modulation is controlled to be higher than the STM feedback cut-off frequency, and the tunneling current signal is then fed into a lock-in amplifier to extract the component having the same modulation frequency. Here, by changing the geometrical shape of the probe from a straight wire to a ring-like shape, the magnetization direction of the probe can be changed from an out-of-plane to an in-plane component. The preparation of the SP-STM tip is still an open field of research, and innovative tip preparation schemes such as using a carbon nanotube as the SP-STM tip are also proposed theoretically [28].

Most of the SP scanning tunneling microscopy experiments require atomically clean tip and sample surfaces. Therefore, the SP scanning tunneling microscopy experiments are mainly conducted in UHV environments. While room-temperature SP scanning tunneling microscopy measurements are demonstrated for a number of systems [12, 29–31], most of the SP scanning tunneling microscopy experiments are conducted at low substrate and tip temperatures. To control the tip magnetization direction, a general SP scanning tunneling microscopy setup may include facilities to apply an internal magnetic field [32–34]. At low temperatures, this can be realized by using superconducting magnets.

10.3 SP Scanning Tunneling Microscopy Applications

10.3.1 *SP Scanning Tunneling Microscopy on Metal Thin Films and Nanostructures*

SP scanning tunneling microscopy has unprecedented power in resolving magnetic structures down to the atomic scale. Different from bulk sensitive measurement techniques such as neutron scattering and vibrating sample magnetometry [35, 36], SP scanning tunneling microscopy directly probes local magnetic information by taking advantage of the tunneling properties. Therefore it is ideal for studying spatially varying nanoscale magnetism on a surface. In this section, we will briefly review some applications of SP scanning tunneling microscopy in resolving nanoscale and atomic-scale magnetic structures in pure metal systems, including ferromagnetic, antiferromagnetic, and vortex states.

One of the first applications of SP scanning tunneling microscopy is resolving the surface magnetism of Cr(001) single crystals [2]. Previous theoretical calculations have predicted a *topological antiferromagnetic* state on this surface [37]. In this magnetic configuration, the magnetization direction alternates for successive terraces separated by single atomic steps. Using a ferromagnetic CrO_2 tip operating in the constant-current mode, Wiesendanger et al. have [2] observed a modulation in the apparent step heights alternating between two height values h_1 and h_2. As explained in the introduction, because of the TMR effect, the tunneling current is enhanced if the tip's and the sample's magnetization directions are parallel and reduced if they are antiparallel. Since the STM here operates in the constant-current mode, the feedback loop adjusts the tip height in order to compensate the differences in tunneling current induced by the TMR effect. Consequently, the tip slightly retracts on the terraces having parallel magnetization to the tip and slightly approaches on the terraces that have antiparallel magnetizations, which results in step height modulations, as illustrated in Fig. 10.3.

It is worth to note that the spin-dependent modulation depends on a variety of parameters. By changing the tunneling bias across the tip–sample junction, the STM probes different electronic states of the sample. Naturally different states could have drastically different spin polarizations, sometimes even with a reversed sign. Furthermore, by changing the tip–sample distance, the tip probes different states (e.g., *s*, *p*, or *d* states) because their decay lengths into the vacuum are not the same. Therefore, the observed modulation amplitude and the sign can depend on the bias as well as on the tip–sample distance.

In the above-described example, the SP component is superimposed onto the spin-averaged topographic signal. A careful analysis is usually required to identify whether the image is SP or not. A judicious way of extracting the spin-dependent signal from the topographic background is imaging in the spectroscopic mode. Here, a small sinusoidal modulation voltage is superimposed onto the tunneling bias, and a lock-in amplifier is used to extract the

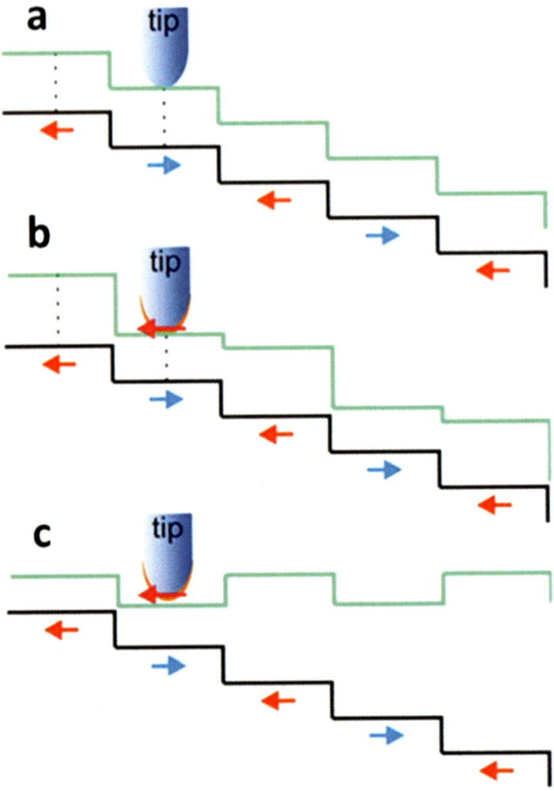

Figure 10.3 Illustration of different imaging modes for SP scanning tunneling microscopy on Cr(001). Black lines denote the topography of the Cr(001) surface terraces. Arrows indicate the spin orientations. Green lines represent the measured line profiles. (a) Topography taken with a nonmagnetic tip, (b) topography taken with a magnetic tip, and (c) dI/dV spectroscopic map taken with a magnetic tip. Schematics are not to scale.

same-frequency response in the tunneling current. The output of the lock-in amplifier is essentially proportional to the first derivative of the tunneling current with respect to the voltage (dI/dV), which in turn is directly proportional to the sample's DOS, assuming a flat DOS for the tip. For the Cr(001) surface, the electronic structure is homogenous across the terraces, and therefore the modulation observed in the spectroscopic mode is mainly originated from the sample magnetization. The dI/dV line profile for the magnetic contrast across the terraces like that of Cr(001) shows just alternating high and low signals without the height information of the terrace (Fig. 10.3). By changing the tunneling bias, the amplitude of the dI/dV square wave could change and sometimes its sign can even reverse. The spectroscopic mode has been widely applied to resolve more complicated magnetic structures on Cr(001) terraces, such as spin frustration near a screw dislocation [38], and on freshly grown Cr spiral terraces [39].

SP scanning tunneling microscopy has proven to be a powerful tool in studying novel nanoscale magnetic phenomena. For example, when an Fe monolayer is grown pseudomorphically on W(001), it exhibits an antiferromagnetic ground state. On the other hand, when an Fe monolayer is grown on W(110), it exhibits a ferromagnetic ground state with an in-plane magnetic anisotropy [40]. When the thickness of the Fe film is increased to two monolayers the anisotropy changes from in plane to out of plane because of the structural relaxations. An example SP scanning tunneling microscopy image of Fe bilayer nanoislands grown on W(110) is shown in Fig. 10.4, where it can be clearly observed that some nanoislands are magnetized along an opposite direction compared to the rest. Minimizing the dipolar stray field energy is the main cause for these opposite magnetizations at zero external magnetic field. When the thickness of the film is increased to three or more layers, the magnetic anisotropy again changes to in plane.

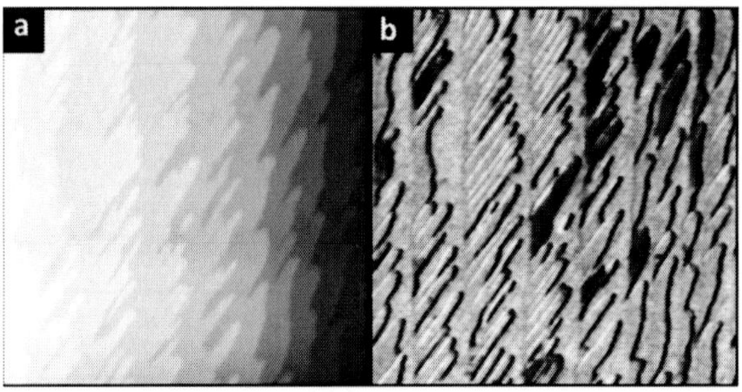

Figure 10.4 SP scanning tunneling microscopy image of an Fe bilayer on W(110) at 35 K. A Cr-coated W tip is used. $V_S = -0.3$V; $I_t = 0.5$ nA. Image size 600 × 600 nm. (a) Topography image and (b) dI/dV map.

Thickness-dependent complex magnetizations are also observed in the case of a thin Fe film grown on a Rh(001) substrate [41].

More exotic magnetic structures have also been observed by SP scanning tunneling microscopy in nanoscale metal systems. Bode et al. have discovered a surface-induced spin spiral state on a Mn monolayer grown pseudomorphically on W(110) [8]. In addition to the interatomic antiferromagnetic coupling, a Dzyaloshinskii–Moriya interaction [42, 43] originating from the spin–orbit interaction is found to induce a long-wavelength modulation, which results in slight canting of the Mn spins away from perfectly antiparallel directions [13, 33]. Because of the nature of this interaction, the spin spiral is also found to be chiral (having a unique sense of rotation) with respect to the surface. The same Dzyaloshinskii–Moriya interaction is also found to be responsible for the formation of the topological protected magnetic states known as *skyrmions* [6, 7], which can be considered as a 2D analog of spin spirals. Skyrmions exhibit a particle-like spin configuration having a topological charge related to the magnetic curvature. Using SP scanning tunneling microscopy, Heinze et al. [7] have directly imaged a skymiron lattice on an Fe layer grown on Ir(111) surface. Recently Romming et al. [6] have demonstrated that individual skyrmions can be written or can be deleted using an SP-STM tip on a PdFe layer grown on Ir(111) lattice. SP scanning tunneling microscopy has also been applied to frustrated spin lattices such as triangular spin lattices. For example, Gao et al. [44] studied the spin structure of one Mn monolayer grown on Ag(111) and observed a so-called Néel structure where the spins are oriented 120° from each other. A similar frustrated spin configuration has also been observed for a Cr/Pd(111) system [45]. On nanoscale ferromagnetic structures, SP scanning tunneling microscopy has been applied to directly probe the internal spin structure of vortex states, where the magnetization direction curls around a core and tilts into an out-of-plane direction in the core region [46]. SP scanning tunneling microscopy has been applied not only to study the spin structures of nanoscale materials and atomic lattices but also to explore the surface electron standing waves [47]. It has been found that the electron standing waves (quantum interference) on a confined magnetic nanostructure can lead to spatial variation in spin polarization [47, 48].

10.3.2 SP Scanning Tunneling Microscopy on Compound Thin Films and Nanostructures

While thin films of pure metal elements could exhibit scientifically intriguing magnetic phenomena, most technologically relevant materials are compound systems containing more than one element (e.g., CoPt, FeMn, etc.). Magnetic imaging of these material systems using SP scanning tunneling microscopy, especially at or beyond room temperature, is generally more demanding than imaging pure metal elements at low temperatures because of the two key reasons: (1) Compound systems usually exhibit an inhomogeneous electronic structure on the surface, making extraction of the magnetic signals difficult, and (2) thermal fluctuations could cause magnetic instabilities at room temperature, leading to a smaller signal-to-noise ratio in the measurements.

Despite these difficulties, several different magnetic compound materials have been successfully investigated with SP scanning tunneling microscopy at room temperature, some of which even achieve atomic-scale spin resolutions [12, 29–31, 49–51]. Using bulk Fe tips, Wiesendanger et al. [49] have reported atomic-resolution magnetic imaging of magnetite (Fe_3O_4), which is a ferrimagnetic material. Here, a superperiodicity equaling 4$\times$ the lattice constant was observed along the Fe atomic rows. This superperiodicity can be explained by a magnetic sensitivity of the tip distinguishing Fe^{3+} from Fe^{2+} ions. Gao et al. [50] have used a CoFeSiB ring as the STM probe and studied the (001) surface of antiferromagnetic $Ni_{50}Mn_{50}$ at room temperature. By applying an oscillating current to the wires wound around the ring, the probe gains an in-plane magnetic sensitivity. Furthermore, by locking onto the oscillation frequency, the extracted signal is originated from the sample magnetism. Here, the atomic lattice is found to be slightly reconstructed on the surface, leading to a noncollinear spin structure that is also energy-dependent. Yamada et al. [51] has also studied the magnetism of Mn–Au alloyed layers by depositing Au onto Mn(001) thin films. Using Fe-coated W tips, the magnetization of each individual Mn–Au layer is resolved, and a layer-wise antiferromagnetic coupling is unveiled for the first three intermixed layers.

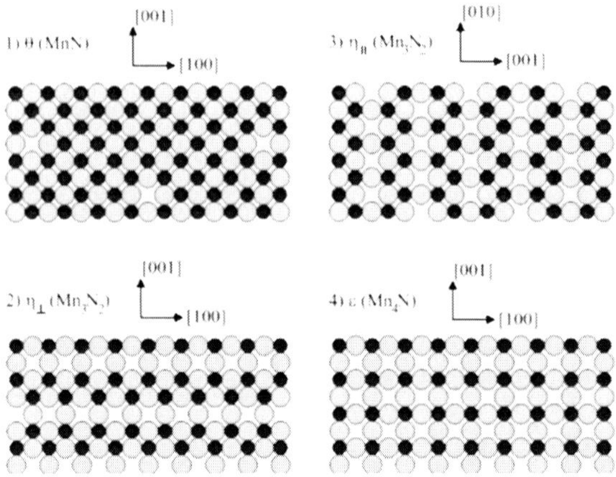

Figure 10.5 Various crystalline phases of manganese nitride. Image courtesy of Arthur R. Smith, Ohio University.

Another interesting compound investigated by using SP scanning tunneling microscopy is Mn_3N_2, which is an antiferromagnet with a high Néel temperature ($\sim$925 K) [12, 52, 53]. Manganese nitride can stabilize in several different crystalline phases when grown by molecular-beam epitaxy (MBE) (Fig. 10.5) [29]. At a 1:1 Mn:N atomic ratio, MnN has a regular rock-salt structure (θ-phase). When the Mn concentration is increased, N vacancies start to form, and at a 3:2 Mn:N ratio, these N vacancies order into repeated planes at every third atomic layer. Depending on the growth parameters, these vacancy planes can be controlled to be either parallel ($\eta_{\parallel}$) or perpendicular ($\eta_{\perp}$) to the film growth direction. A further increase of the Mn:N ratio to 4:1 yields the ϵ-phase where the N vacancy planes now repeat for every second atomic layer. Among these phases, θ- and η-phases are both antiferromagnetic with Néel temperatures higher than room temperature, whereas the ϵ-phase is ferrimagnetic.

We will now focus on SP scanning tunneling microscopy studies of η-Mn_3N_2 [29]. As shown in Fig. 10.6, η-Mn_3N_2 has a regular rock-salt structure except for every third layer, which has voids of N atoms. The bulk spin structure determined by neutron scattering is known to exhibit a layer-wide antiferromagnetism. The Mn spins are ferromagnetically coupled (parallel) within each (001) layer, while

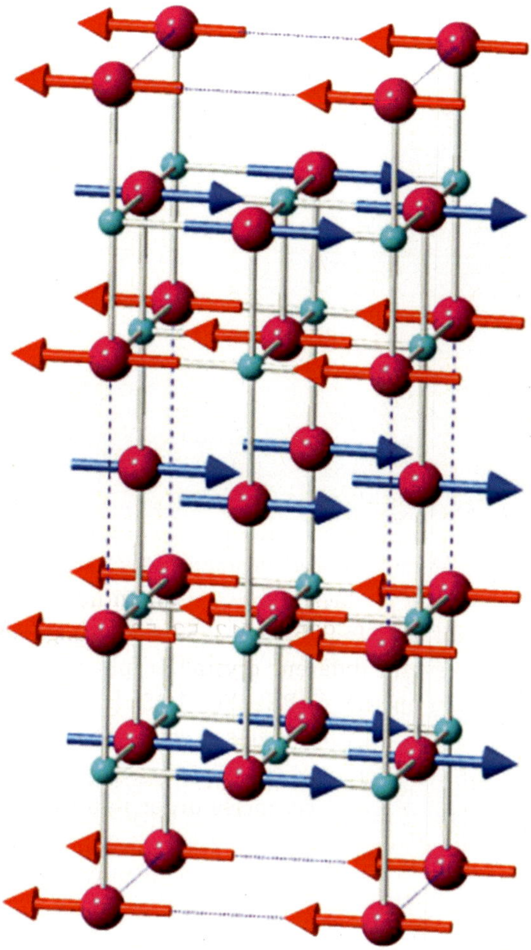

Figure 10.6 Crystal structure of η-Mn_3N_2. Purple atoms represent Mn and cyan atoms represent N. Arrows indicate the spin direction of each Mn atom.

their orientations alternate for every successive layer. For the $\eta_{\parallel}$-phase where the surface is (010) oriented, these N vacancy planes intersect with the surface, forming N vacancy rows. The nearest row–row distance equals $\frac{c}{2} = 6.07$ Å. A representative scanning tunneling microscopy image of this surface taken with nonmagnetic W tips is shown in Fig. 10.7. The observed atomic rows here directly correspond to the N vacancy rows and have a row–row spacing of

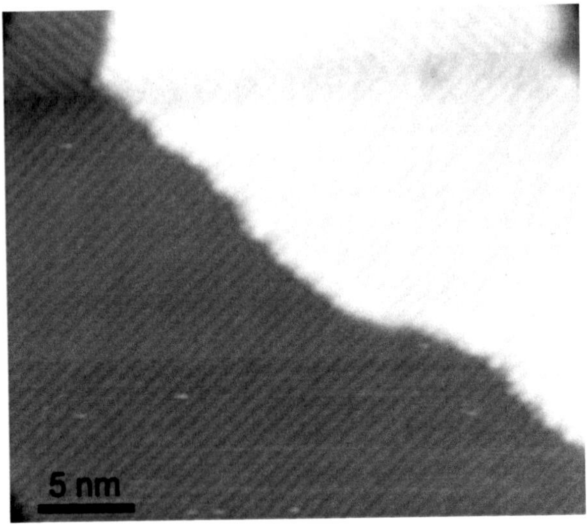

Figure 10.7 A representative scanning tunneling microscopy image of $\eta_{\|}$-Mn_3N_2 surface taken with a nonmagnetic tip. $V = -0.3\mathrm{V}; I = 0.12\mathrm{nA}$ Bright rows correspond to the N vacancy rows on the surface. The 90° turn in the row directions indicates a domain boundary that extends through different layers.

6.07 Å. At the negative biases, the local electron densities of the Mn atoms, which are not in the N vacancy rows, are lower than the ones in the N vacancy rows and are difficult to resolve in scanning tunneling microscopy images unless an extremely sharp tip is used.

By using a magnetically coated tip, magnetic sensitivity can be achieved, in addition to atomic resolution. When the tip's magnetization direction is parallel with the local Mn atom's magnetization direction, tunneling current is enhanced, as explained before. This results in a modulation in the apparent heights of the atomic rows. From the Fig. 10.6, we can easily see that the adjacent N vacancy rows have opposite spin orientations. Therefore the modulation period (c) will have a double periodicity of the structural period ($c/2$). An example of such magnetic image is shown in Fig. 10.8. In addition to the normal row resolution, upon close inspection, one can find that the rows' heights now alternate between high and low (brighter and darker in the image). A

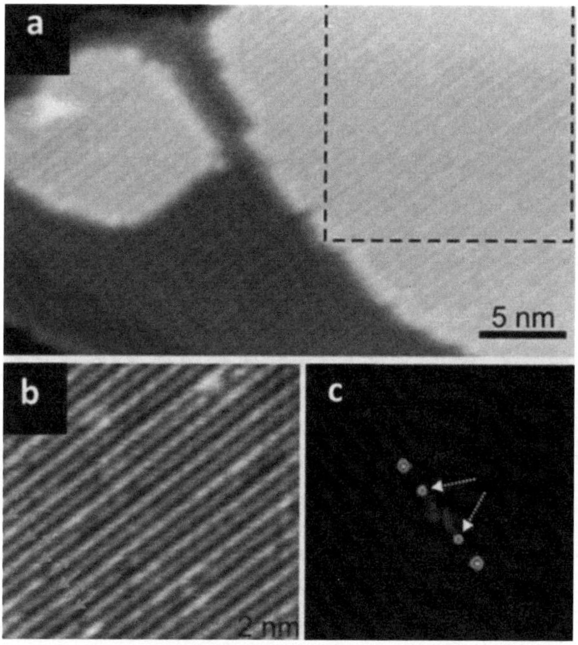

Figure 10.8 A representative SP scanning tunneling microscopy image of $\eta_{\parallel}$-Mn_3N_2 surface taken with a Fe-coated W tip. $V = -0.3$ V; $I = 0.2$ nA A close-up view of the boxed area in (a) is shown in (b) where the alternating brighter and darker atomic rows are more apparent. A fast Fourier transform of image (b) is shown in (c). In addition to the reciprocal lattice points corresponding to the $c/2$ atomic lattice, additional spots (pointed by arrows) indicate spin periodicities.

close-up to the boxed area is shown in Fig. 10.8b where this effect is more apparent.

Fast Fourier transform of the real-space image clearly reveals two weaker k-spots (indicated by arrows) in addition to the regular reciprocal lattice points, which are the brighter spots. This double-period modulation obtained by using a magnetic tip is originated from the alternating spin orientations of the Mn atoms in these rows, and it is a direct proof of the antiferromagnetic nature of the surface. The exact modulation amplitude depends on a variety of factors, including the spin DOS, the tip–sample separation, and the tip's spin polarization [30]. By changing the growth parameters, one

Figure 10.9 Scanning tunneling microscopy image of $\eta_\perp$-Mn_3N_2 showing many nanopyramids formed on the surface [12].

can also obtain the $\eta_\perp$-phase where the N vacancy planes are now parallel instead of perpendicular to the surface. The surface of $\eta_\perp$-Mn_3N_2 is found to consist of many nanopyramids having lateral sizes ranging from 20 to 100 nm (Fig. 10.9). Using a soft magnetic tip whose magnetization direction can be readily controlled, the 3D spin structure of these surface nanopyramids can be mapped out at room temperature [12].

As shown in Fig. 10.9 these nanopyramids have flat atomic terraces that are separated by single-atomic steps (roughly $c/6$). Notice that these atomic layers are not electronically equivalent because of the different local environments. In bulk lattices (see Fig. 10.6), the N vacancy planes (referred to as the Mn layers) have a different LDOS from that of the N-filled planes (referred to as the MnN layers). At the film surface, the electronic equivalence of the two adjacent MnN layers is further broken due to the lack of inversion symmetry. The different electronic structure of these atomic layers is difficult to distinguish in the topographic channel but it is manifested in the dI/dV channel, which is directly

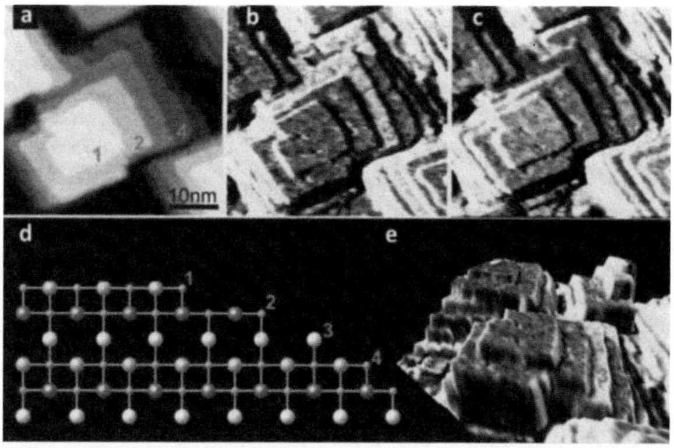

Figure 10.10 Scanning tunneling microscopy images of $\eta_{\perp}$-Mn_3N_2 showing (a) topography, (b) dI/dV map taken at $V = -0.3$V, and (c) dI/dV map taken at $V = 0.2$V Images are taken with a nonmagnetic tip at room temperature. (d) Assignment of different atomic terraces to the different atomic layers in the lattice. 1, 2, and 4 correspond to the MnN layer, while 3 corresponds to the Mn layer, respectively. (e) 3D rendered composite image with colors representing dI/dV signals [12].

proportional to the LDOS. As shown in Fig. 10.10, at −0.3 V bias, the two MnN layers exhibit similar contrasts (terraces 1 and 2 in Fig. 10.10b), whereas the Mn layer has a higher dI/dV signal (terrace 3 in Fig. 10.10b). This contrast reverses at +0.2 V where the Mn layer now has a lower dI/dV signal than the MnN layers. The electronic structure variation in the MnN layers induced by the surface symmetry breaking is only noticeable at positive biases. As a consequence, terrace 1 is slightly brighter than terrace 2 in Fig. 10.10c. The assignment of these terraces to the different atomic layers is shown in Fig. 10.10d. A smart way to present these data is to render the topography in 3D while texturing it on the dI/dV map. An example is shown in Fig. 10.10e, where the topography and the dI/dV map taken at −0.3 V are now superimposed to form a single composite image.

According to the bulk spin structure shown in Fig. 10.6, each adjacent MnN layers is expected to have opposite spin orientations, and both should lie within the surface plane. Therefore by using an in-plane sensitive tip, it should exhibit a modulation in the SP-STS dI/dV signal of these terraces. In Fig. 10.11, a new contrast

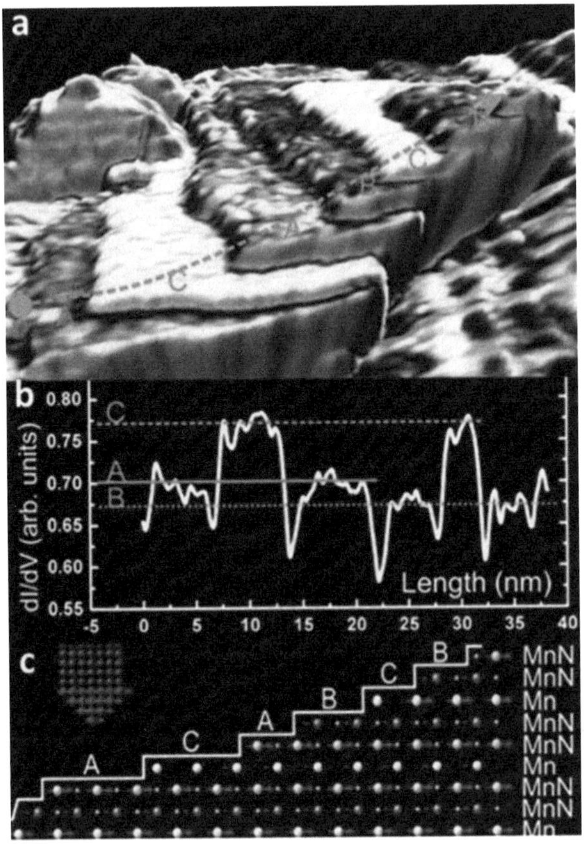

Figure 10.11 (a) 3D composite image taken with an Fe-coated W tip. $V = -0.3\text{V}$; $I = 0.1\text{nA}$. (b) dI/dV line profile going up the terraces, showing a sequence of *ACABCA*. (c) Atomic and spin model of the nanopyramids as expected from the bulk spin structure [12].

emerges among the adjacent MnN layers at −0.3 V when using an Fe-coated W tip, which is sensitive to in-plane magnetization. Furthermore, the contrast sequence obtained when using an Fe-coated tip is different from the regular electronic sequence obtained when using a nonmagnetic W tip. The nonmagnetic electronic sequence at −0.3 V can be labeled as *AACAAC*... (Fig. 10.10), where *A* corresponds to the MnN layers and *C* corresponds to the Mn layers, respectively. Strikingly, when a magnetic tip is used, the contrast sequence now converts to *ACABCB...* (Fig. 10.11). The letter can only be explained by a spin sequence. As shown in Fig. 10.11c, the MnN layer whose spins are parallel with the tip's spin exhibits an enhanced dI/dV signal, while that whose spins are antiparallel exhibits a reduced dI/dV signal. It is amenable from the expected spin structure that the spin-sensitive contrast sequence to be *ACABCB...* and it is consistent with the observed sequence.

While the MnN layers have been magnetically resolved and are consistent with the expected layer-wide antiferromagnetic spin structure, a magnetic contrast among the Mn layers has not been obtained. To explore the spin structure of these Mn layers, one is required to investigate using different bias voltages because the spin polarization can be strongly energy dependent. However, the spin contrast is not obtained for all the biases between -1 and 1 V. This observation suggests that either these layers are magnetically inactive or their spins have reoriented into different directions.

The custom-built SP scanning tunneling microscopy setup used for this experiment allows switching the magnetization direction of the tip from in-plane to out-of-plane directions by applying a $\sim$0.4 T magnetic field parallel to the tip axis during tunneling [34]. A comparison of the SP scanning tunneling microscopy images taken with an in-plane tip and an out-of-plane tip (the same tip with only different magnetization directions) is shown in Fig. 10.12. Figure 10.12a is taken without the external field, while Fig. 10.12b is acquired with an applied external magnetic field. Comparison of these two images reveals that the in-plane contrast among the MnN layers vanishes under an applied magnetic field, while the electronic contrast between the Mn and the MnN layers remains the same. Thus the magnetic origin of the contrast observed among MnN layers (Avs.B) is apparent. When the magnetic field rotates the tip's magnetization direction to an out-of-plane, the tip loses magnetic sensitivity to the in-plane magnetized MnN layers, and the A–B contrast is lost, as expected.

Interestingly, with the out-of-plane tip, a new contrast emerges among the Mn layers (Fig. 10.13). This contrast obtained by an

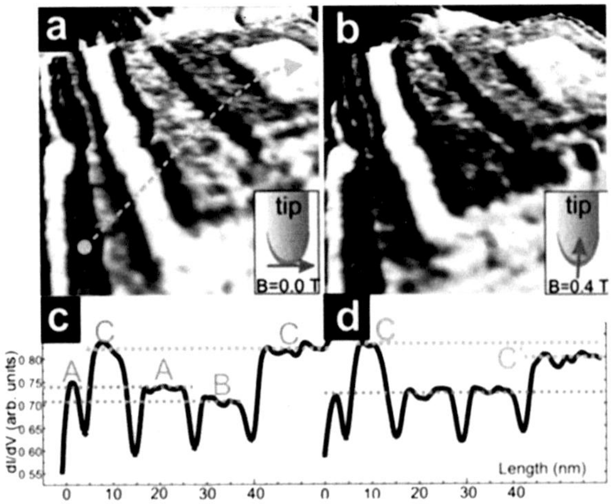

Figure 10.12 SP scanning tunneling microscopy images of the same nanopyramid taken (a) without and (b) with the external magnetic field applied. (c) and (d) are the dI/dV line profiles of (a) and (b), respectively [12].

out-of-plane tip indicates an out-of-plane magnetization of the exposed Mn layers, which is contrary to the expectations based on bulk spin structure. Since the electronic contrast remains unchanged during the rotation of the tip's magnetization direction, one can remove this contrast by subtracting the dI/dV signal of Fig. 10.12a from that of Fig. 10.12b. By doing so, a pure magnetic signal can be extracted, and the result is shown in Fig. 10.13b, where the difference in the dI/dV signal is superimposed onto the 3D rendered topographic image. Notice that the in-plane contrast and the out-of-plane contrast are both reproduced by subtraction. Thus, these measurements enable to derive the 3D spin structure of the nanopyramid shown in Fig. 10.13c. The MnN layers here have alternating in-plane magnetizations consistent with the bulk spin structure, while the magnetizations of the exposed Mn layers have reoriented into alternating out-of-plane directions. Moreover, the Mn atoms in the exposed Mn layers have been found to be tilted inward and form tetramer clusters, thereby drastically changing their atomic coordination compared to the bulk-case. Such effect might be induced by surface reconstructions.

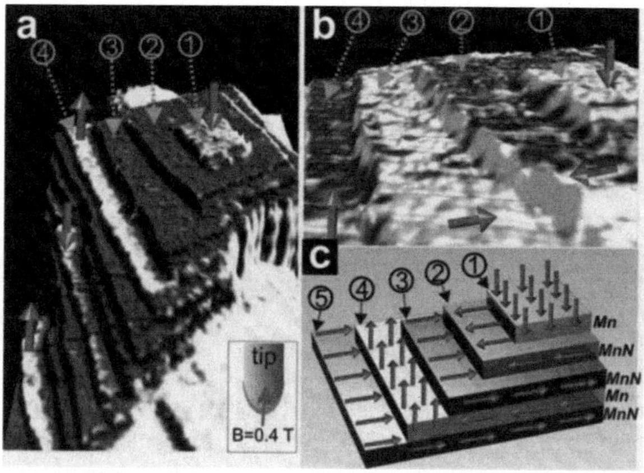

Figure 10.13 (a) The same image as Fig. 10.12b but with the contrast levels readjusted to focus on the Mn layers. (b) The difference image between Figs. 10.12b and 10.12a. (c) A schematic of the derived 3D spin structure of the nanopyramid [12].

10.3.3 *Imaging the Spin Direction of Individual Atoms*

SP scanning tunneling microscopy is also a very useful tool to image the spin contrast of individual atoms [10, 13, 54–56]. Imaging of spin directions in individual atoms have been demonstrated by Serrate et al. [10] using Co atoms on a Mn layer grown on a W(110) surface. The monolayer islands of Mn formed on a W(110) surface can exhibit an antiferromagnetic spin spiral due to the Dzyaloshinskii–Moriya interaction [8]. This antiferromagnetic spin spiral propagates along the [1-10] surface direction with an angle of $\sim 170^\circ$ between magnetic moments of adjacent atomic rows. When the images are taken with a magnetic tip at $\sim$10 K substrate temperature, stripes along the [001] direction of the Mn layer produced by the spin spiral can be observed (Fig. 10.14a). These stripes are originated from the antiferromagnetic spins where the higher (brighter) parts of the stripes are produced by the Mn atomic rows with their spin directions toward the tip spin, while the lower (darker) parts have spins toward the opposite direction of the tip spin. This height contrast vanishes at approximately every 6 nm along the [1–10] direction where the local Mn moments are in plane. To maintain the tip spin direction, an internal magnetic field

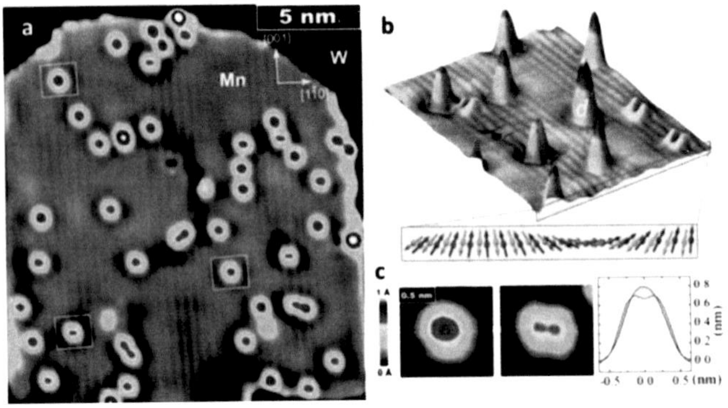

Figure 10.14 (a) Co atoms adsorbed on a Mn monolayer island on W(110). The stripes on the surface at the background are caused by the spin spiral. (b) A 3D view of adsorbed atoms together with a drawing showing a spin spiral lattice. (c) A close-up view of two different Co atoms' appearance and corresponding line profiles [10].

of 2.5 T along the out-of-plane direction is applied to the sample. This applied magnetic field is chosen so that it does not alter the antiferromagnetic state of the Mn layer.

It is known that Co atoms ferromagnetically coupled to the surface magnetic moment at the adsorption sites [10]. Therefore, the magnetization directions of Co atoms will not only vary along [1–10] direction if one moves the atoms across the surface but also will change to the opposite moment at the neighboring sites. In accord with this, SP scanning tunneling microscopy images of the Co atoms on a Mn monolayer island show different heights and shapes (Fig. 10.14b). The Co atom heights are higher when they are located on bright Mn atom rows and lower when they are on darker Mn atom rows. In the letter case, the atoms also appear as double-lobe features (Fig. 10.14c). At the sites where Mn moments are perpendicular to the tip magnetization the Co atoms have an intermediate height. In accordance with the experimental SP scanning tunneling microscopy images of Co atom shapes, the calculated LDOS in the vicinity of the Fermi energy reveals an almost circular shape of Co atom for the majority channel, while the minority channel shows two lobes along the [1–10] direction. In the majority channel, the Co atom has an LDOS composed of s, p_z,

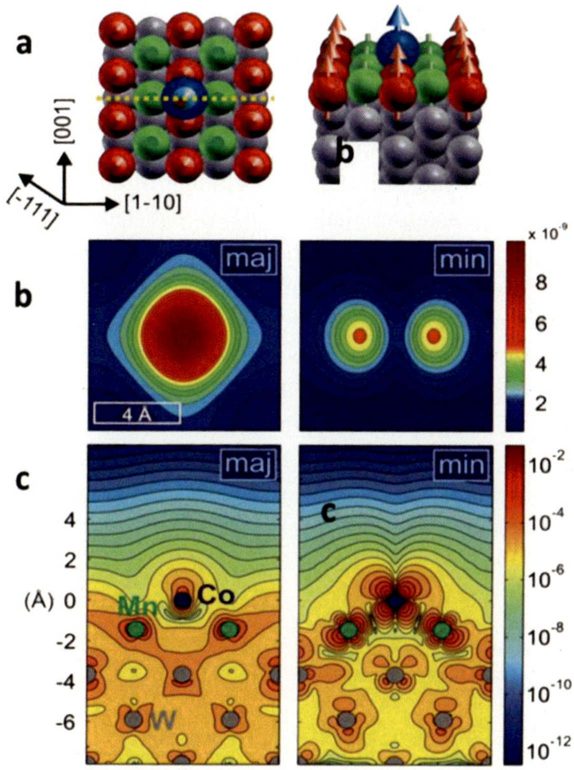

Figure 10.15 Calculated spin-resolved local density of states of a Co atom on Mn/W(110) at E_F. (a) Top and perspective views of the unit cell used in the density functional theory (DFT) calculations. Red and green spheres represent Mn atoms with magnetization parallel and antiparallel to the Co atom (blue), respectively. (b) LDOS (given in $e/(a_0)^3$, where a_0 is the Bohr radius) in the energy interval between E_F and E_F+ 10 meV for majority (maj) and minority (min) electrons in a plane parallel to the surface 5.2 Å above the Co atom. (c) Same as (b) for a cross section perpendicular to the surface along the dashed line in (a). The positions of Co, Mn, and W atoms are marked for clarity [10].

and d_z^2 orbitals. For the minority channel, the LDOS of Co atom is mainly contributed from a d_{xz} state, which appears split at the Co atom center (Fig. 10.15). Thus, the appearance of round and split shapes Co atoms are due to their spin-up and spin-down states on the Mn monolayer. This is a remarkable finding, enabling a direct visualization of the spin-up and spin-down states at the single-atom level.

In addition, the spin spiral substrate of Mn monolayer has varying spin directions, and the spins of Co atoms adsorbed on different adsorption sites of this surface should exhibit accordingly. This is reflected in an artificial atomic chain formed by six individual Co atoms that are positioned along a [1–10] surface direction. Here, the Co atom chain covers half of the spin spiral period. The Co atoms in the atomic chain exhibit a gradual transition from a spin-down appearance, that is, the split shape, to a spin-up appearance, that is, the round shape (Fig. 10.16), and their heights also vary gradually. As shown in Fig. 10.16a,b, by switching the internal magnetic field from +2.5 T to −2.5 T, the tip spin direction is reversed, and hence, the shapes of the Co atoms in the other half of the spin spiral period can be imaged. Here, the Co spin direction can be determined directly from the height profile because of the cos θ dependence of the varying spin along the chain (Fig. 10.16d).

10.3.4 *Spin Friction*

Manipulation of atoms on a surface can be not only used to construct atomic structures, as shown above, but also used to explore atomic-scale tribology. During a lateral manipulation of a silver atom on a Ag(111) surface with the STM tip, the atom movements resembling smooth-sliding and stick-slip styles have been found [57]. If one combines the scanning tunneling microscopy atom manipulation technique with SP scanning tunneling microscopy, it is possible to explore contribution of spin degrees of freedom in frictional phenomena at the atomic scale. This has been demonstrated by Wolter et al. [11] by laterally manipulating a Co atom with a magnetic SP-STM tip across a spin spiral surface [8, 10] formed by a Mn monolayer on a W(110) substrate (Fig. 10.17a). The Co atom manipulation here involves an attractive tip atom force originated from the overlapping of the electronic wave functions between the tip apex atom and the manipulated atom [57]. However, because both the tip and the manipulated atom are formed by magnetic atoms, the magnetic exchange interactions also exist between them.

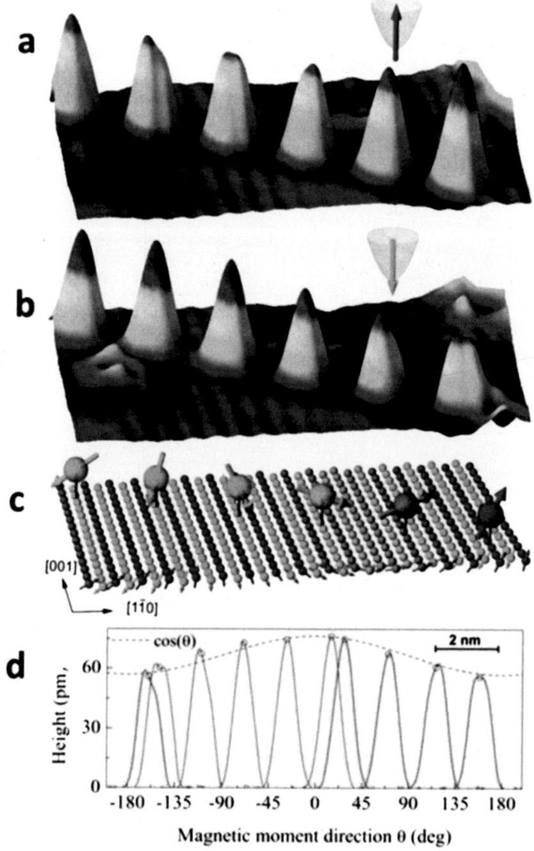

Figure 10.16 Perspective views of constant-current images ($U = -10$ mV, $I = 2$ nA, Fe-coated W tip) taken in (a) $B = +2.5$ T and (b) $B = -2.5$ T. Raw data has been smoothed with a Gaussian in the [001] direction to reduce noise. (c) The atoms are equally spaced with an angle of $\sim 42^\circ$ between adjacent Co moments. (d) Line profiles of the above data taken along the [1–10] direction and positioned as a function of θ. The apparent height closely follows a $\cos\theta$ law [10].

This provides additional information about their magnetic states and consequently the contribution of spin in frictional processes.

Because the Co atom is moved in a 2D surface region using the constant-current mode, the resulting tip height signal (which is referred to as manipulation signal) can be recorded as a 2D map.

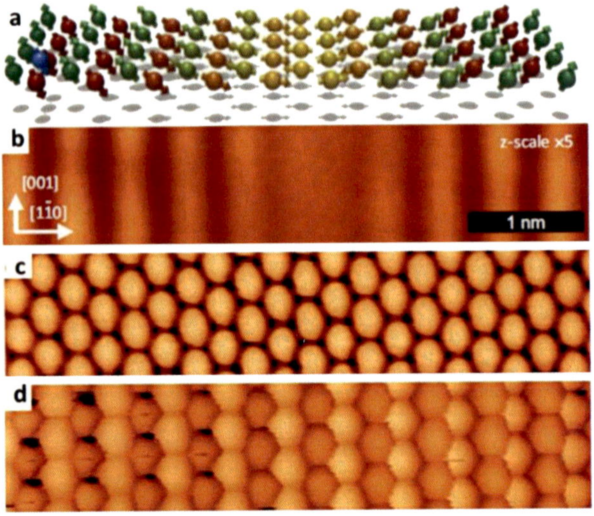

Figure 10.17 Illustration of the spin spiral surface and the manipulated atom (blue) (a) and SP scanning tunneling microscopy image of the spin spiral (b). The manipulated images of the Co atom on the spin spiral surface with a W tip (c) and an Fe tip (d) [11].

In a normal SP scanning tunneling microscopy image shown in Fig. 10.17b, the spin spiral surface is composed of stripes with varying intensities. As discussed above, the higher and lower contrast forming the stripes are generated by the parallel and antiparallel magnetization directions between the tip and the sample. Because of the antiferromagnetic spin spiral, the magnetization directions between the tip and the sample vary with the location. When the magnetization direction of Mn is closed to perpendicular to the tip magnetization direction, the corrugation height of the stripes vanishes due to a cos θ dependent on the tip and sample magnetization directions [10].

To demonstrate the magnetic effect, first a Co atom is manipulated with a nonmagnetic W tip, and the image shows the atomic lattice of the surface (Fig. 10.17c) [11]. However, the atomic features observed in Fig. 10.17c are not originated from the surface Mn atom. These features are formed by a single Co atom during the manipulation, which moves in a rest-hop cycle across the surface in a 2D manipulation area [11, 58]. This mechanism can be explained as follows: When the Co atom is adsorbed on a surface hollow site, it is in a static position (at rest), and the tip passes over the atom, thereby tracing the top part of the Co atom contour, producing the shape of the Co atom in the manipulation image. Then when the tip reaches at the end part of the atomic contour the atom rapidly moves to the next surface site and positions in front of the tip. Because the Co atom movement is much faster than the tip scanning speed, the tip does not record the hoping traces. Now the tip starts the process of following the atomic contour of the Co atom in the new position again. Repeating this process enables mapping of the entire surface. Since the Co atom is temporarily at rest in every hollow sites of this 2D manipulation area, the image has the same structural symmetry as the Mn lattice underneath. When the magnetic effect is absent, that is, in the scanning tunneling microscopy topographic mode, the Mn atoms should appear an equal size with the same height within the surface plane, as shown in this Co atom manipulation image.

When the Co atom manipulation image is recorded with a magnetic tip, then the appearance of the surface atomic structures are drastically different (Fig. 10.17d). The surface features of manipulation image include corrugations with one of the atomic row appears higher than the other. Moreover, the sizes of the atomic shapes are also different. In some areas, they appear wider, while in other locations, they appear narrower. This apparent change in the atom shapes and height can be directly tied to the effect of the spin, which reflects the tip height and therefore the SP tunneling current intensity during the lateral movement of the Co atom. It also contains lateral force information as well, which now also has contributions from spin exchange interactions.

The observed atomic corrugation can be simply explained as follows: The brighter and darker appearance (or higher and lower heights) that forms corrugations (or stripes) can be understood with the increase or decrease in tunneling current that alerts the tip to retract or approach toward the atom during the manipulation by the STM feedback because of the constant-current tunneling mode. Since the topographic structures of the Mn atoms in the spin spiral are equal and located in the same surface plane, this change in height when a magnetic tip is used has to be related to the magnetic origin of the surface atoms. For the different sizes of the atoms, that is, wider or narrower appearances of the atoms, it can be explained with the time spent by the Co atom at each adsorption site on the surface. When the Co atom stays for a shorter time at an adsorption site, then the tip will trace only a small fraction of the atomic contour and it will appear narrower. But if the Co atom stays longer at the site, then the tip will trace more of its contour and it will now appear wider. As mentioned in previous sections, one can explain this effect as caused by the alignment of the tip and Co atom magnetization directions toward parallel and antiparallel configuration with a varying angle θ at different locations due to the spin spiral surface.

The mechanism of the magnetic exchange interaction occurring at the junction of the tip, the manipulated atom and the surface can be explained as follows: When a ferromagnetic coupling between the tip, the Co atom, and the surface, that is, their spin directions, is parallel, then the system is in an energy minimum. When the tip and surface magnetization directions are antiparallel, then the Co atom is in a frustrated spin configuration. In this case, the Co atom magnetization will be canted toward the surface magnetization for an increased tip height (Fig. 10.18a), and canted toward the tip magnetization for a reduced tip height (Fig. 10.18b). When the Co atom is manipulated with a nonmagnetic tip, the magnitude of the

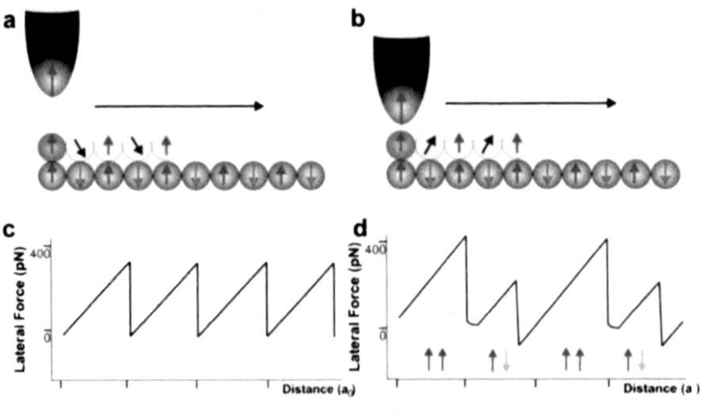

Figure 10.18 (a) A far SP-STM tip and (b) a close SP-STM tip to the manipulated atom on top of an antiferromagnetic surface. Lateral force as a function of the atomic lattice site during manipulation with a nonmagnetic tip (c) and a magnetic tip (d).

lateral force component as a function of lateral distance is illustrated in Fig. 10.18c. When the Co atom is at rest on a surface hollow site during the manipulation, the continuous tip scanning causes the lateral force to gradually increase, producing a linear rise in the force slope. When the lateral force component overcomes the surface retaining forces the Co atom moves to the next adsorption site on the surface. At this point, the lateral force component abruptly reduces to a minimum value. Repeating this process of the rest-hop cycle movement, the resulting lateral force across the surface during the Co atom manipulation appears in a symmetric saw-tooth shape.

When the Co atom is manipulated with a magnetic tip, then the situation is different. The Co atom in the frustrated spin configuration at the adsorption site where the tip and surface magnetizations are antiparallel encounters a reduced diffusion barrier to move to the next site. As a result, the atom stays a shorter period then the parallel spin configuration. On the other hand, at the sites with the tip–sample spin parallel configuration, the atom stays longer period because the barrier for diffusion of Co atom increases. In a way, the Co atom prefers to sit there for a longer period. As a result, the lateral force component to move the Co atom to the next site also increases. The Monte Carlo simulation of this Co atom

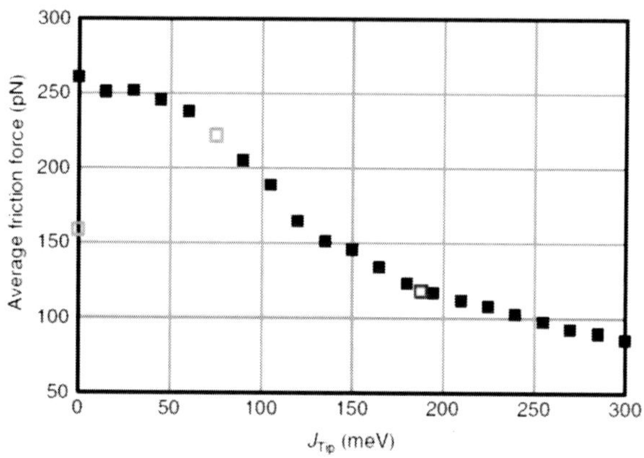

Figure 10.19 Average friction force of a magnetic atom on a row-wise antiferromagnetic surface as a function of exchange interaction strength [11].

manipulation reveals that the frictional force can increase from ~80 pN to ~260 pN (Fig. 10.19), depending on the magnetic exchange interaction strength. This clearly demonstrates for the first time that atomic spin can greatly contribute to the frictional phenomena.

10.4 Summary

The above examples clearly show that the magnetism at reduced dimensions and nanoscale structures can be very different from their bulk counter parts. At reduced dimension and symmetries, additional spin-dependent couplings such as spin–orbit, Rudel–Kittel–Kasuya–Yosida, and Dzyaloshinskii–Moriya interactions start to play important roles and can often lead to the discovery of novel magnetic phenomena. The unexpected 3D spin structure observed on the Mn_3N_2 nanopyramids is a good example of how magnetism can behave drastically different when material sizes are reduced and symmetries are broken. At the atomic scale, the spin-dependent friction on a spin spiral surface clearly is an example how novel phenomena can be observed at the atomic scale. A recent extension of SP scanning tunneling microscopy application to magnetic molecules adsorbed on surfaces [59, 60] also highlights the fact that this technique is still opening many new research directions. Future technological applications such as high-density nonvolatile memories will inevitably rely on detailed understanding of magnetism in nanoscale materials. By providing the ultimate resolution on atomic and spin structures, SP scanning tunneling microscopy has been and will continue to be a powerful tool to explore novel magnetic phenomena locally and will aid in achieving the goals of spintronic applications.

Acknowledgments

Use of the Center for Nanoscale Materials is supported by the US Department of Energy, Office of Science, Office of Basic Energy Sciences, under contract no. DE-AC02-06CH11357.

References

1. G. Binnig and H. Rohrer, Scanning tunneling microscopy, *Helv. Phys. Acta*, **55**, 726–735 (1982).
2. R. Wiesendanger, H. J. Güntherodt, G. Güntherodt, R. J. Gambino, and R. Ruf, Observation of vacuum tunneling of spin-polarized electrons with the scanning tunneling microscope, *Phys. Rev. Lett.*, **65**, 247 (1990).
3. S. F. Alvarado and P. Renaud, Observation of spin-polarized-electron tunneling from a ferromagnet into GaAs, *Phys. Rev. Lett.*, **68**, 1387–1390 (1992).
4. M. Julliere, Tunneling between ferromagnetic-films, *Phys. Lett. A*, **54**, 225–226 (1975).
5. J. Tersoff and D. R. Hamann, Theory of the scanning tunneling microscope, *Phys. Rev.* B, **31**, 805–813 (1985).
6. N. Romming, C. Hanneken, M. Menzel, J. E. Bickel, B. Wolter, K. von Bergmann, A. Kubetzka, R. Wiesendanger, Writing and deleting single magnetic skyrmions, *Science*, **341**, 636–639 (2013).
7. S. Heinze, K. von Bergmann, M. Menzel, J. Brede, A. Kubetzka, R. Wiesendanger, G. Bihlmayer, and S. Blügel, Spontaneous atomic-scale magnetic skyrmion lattice in two dimensions, *Nat. Phys.*, **7**, 713–718 (2011).
8. M. Bode, M. Heide, K. von Bergmann, P. Ferriani, S. Heinze, G. Bihlmayer, A. Kubetzka, O. Pietzsch, S. Blügel, and R. Wiesendanger, Chiral magnetic order at surfaces driven by inversion asymmetry, *Nature*, **447**, 190 (2007).
9. P. Ferriani, K. von Bergmann, E. Y. Vedmedenko, S. Heinze, M. Bode, M. Heide, G. Bihlmayer, S. Blugel, and R. Wiesendanger, Atomic-scale spin spiral with a unique rotational sense: Mn monolayer on W(001), *Phys. Rev. Lett.*, **101**, 027201 (2008).
10. D. Serrate, P. Ferriani, Y. Yoshida, S.-W. Hla, M. Menzel, K. von Bergmann, S. Heinze, A. Kubetzka, and R. Wiesendanger, Imaging and manipulating the spin direction of individual atoms, *Nat. Nanotechnol.*, **5**, 350–353 (2010).
11. B. Wolter, Y. Yoshida, A. Kubetzka, S. W. Hla, K. von Bergmann, and R. Wiesendanger, Spin friction observed on the atomic scale, *Phys. Rev. Lett.*, **109**, 116102 (2012).
12. K. K. Wang and A. R. Smith, Three-dimensional spin mapping of antiferromagnetic nanopyramids having spatially alternating surface anisotropy at room temperature, *Nano Lett.*, **12**, 5443–5447 (2012).

13. M. Heide, G. Bihlmayer, and S. Blügel, Dzyaloshinskii-Moriya interaction accounting for the orientation of magnetic domains in ultrathin films: Fe/W(110), *Phys. Rev. B*, **78**, 140403 (2008).
14. F. Meier, L. H. Zhou, J. Wiebe, and R. Wiesendanger, Revealing magnetic interactions from single-atom magnetization curves, *Science*, **320**, 82–86 (2008).
15. A. Li Bassi, C. S. Casari, D. Cattaneo, F. Donati, S. Foglio, M. Passoni, C. E. Bottani, P. Biagioni, A. Brambilla, M. Finazzi, F. Ciccacci and L. Duo, Bulk Cr tips for scanning tunneling microscopy and spin-polarized scanning tunneling microscopy, *Appl. Phys. Lett.*, **91**, 173120 (2007).
16. A. Schlenhoff, S. Krause, G. Herzog and R. Wiesendanger, Bulk Cr tips with full spatial magnetic sensitivity for spin-polarized scanning tunneling microscopy, *Appl. Phys. Lett.*, **97**, 083104 (2010).
17. M. Bode, Spin-polarized scanning tunnelling microscopy, *Rep. Prog. Phys.*, **66**, 523–582 (2003).
18. J. Prokop, A. Kukunin, and H. J. Elmers, Spin-polarized scanning tunneling microscopy and spectroscopy of ultrathin Fe/Mo(110) films using W/Au/Co tips, *Phys. Rev. B*, **73**, 014428 (2006).
19. A. Kubetzka, M. Bode, O. Pietzsch, and R. Wiesendanger, Spin-polarized scanning tunneling microscopy with antiferromagnetic probe tips, *Phys. Rev. Lett.*, **88**, 057201 (2002).
20. G. Rodary, J.-C. Girard, L. Largeau, C. David, O. Mauguin, and Z.-Z. Wang, Atomic structure of tip apex for spin-polarized scanning tunneling microscopy, *Appl. Phys. Lett.*, **98**, 082505 (2011).
21. L. Berbil-Bautista, S. Krause, M. Bode and R. Wiesendanger, Spin-polarized scanning tunneling microscopy and spectroscopy of ferromagnetic Dy(0001)/W(110) films, *Phys. Rev. B*, **76**, 064411 (2007).
22. T. K. Yamada, M. M. J. Bischoff, T. Mizoguchi and H. Van Kempen, Use of voltage pulses to detect spin-polarized tunneling, *Appl. Phys. Lett.*, **82**, 1437–1439 (2003).
23. C. F. Hirjibehedin, C. Y. Lin, A. F. Otte, M. Ternes, C. P. Lutz, B. A. Jones and A. J. Heinrich, Large magnetic anisotropy of a single atomic spin embedded in a surface molecular network, *Science*, **317**, 1199–1203 (2007).
24. W. Wulfhekel and J. Kirschner, Spin-polarized scanning tunneling microscopy on ferromagnets, *Appl. Phys. Lett.*, **75**, 1944–1946 (1999).
25. U. Schlickum, W. Wulfhekel, and J. Kirschner, Spin-polarized scanning tunneling microscope for imaging the in-plane magnetization, *Appl. Phys. Lett.*, **83**, 2016–2018 (2003).

26. C. L. Gao, U. Schlickum, W. Wulfhekel and J. Kirschner, Mapping the surface spin structure of large unit cells: reconstructed Mn films on Fe(001) *Phys. Rev. Lett.*, **98**, 107203 (2007).

27. C.-B. Wu, J. Song, and W. Kuch, Spin-polarized scanning tunneling microscopy study of Mn/Co/Cu(001) using a bulk Fe ring probe, *Appl. Phys. Lett.*, **101**, 012404 (2012).

28. X. Zou, G. Zhou, J. Li, B. Yan, J. Wu, B.-L. Gu, and W. Duan, Preparing spin-polarized scanning tunneling microscope probes on capped carbon nanotubes by Fe doping: a first-principles study, *Appl. Phys. Lett.*, **94**, 193106 (2009).

29. H. Q. Yang, A. R. Smith, M. Prikhodko, and W. R. L. Lambrecht, Atomic-scale spin-polarized scanning tunneling microscopy applied to Mn3N2(010) *Phys. Rev. Lett.*, **89**, 226101 (2002).

30. R. Yang, H. Q. Yang, A. R. Smith, A. Dick, and J. Neugebauer, Energy-dependent contrast in atomic-scale spin-polarized scanning tunneling microscopy of Mn_3N_2 (010): experiment and first-principles theory, *Phys. Rev. B*, **74**, 115409 (2006).

31. I. Altfeder, W. Yi, and V. Narayanamurti, Spin-polarized scanning tunneling microscopy of the room-temperature antiferromagnet *c*-FeSi, *Phys. Rev. B*, **87**, 020403R (2013).

32. J. Wiebe, A. Wachowiak, F. Meier, D. Haude, T. Foster, M. Morgenstern, and R. Wiesendanger, A 300 mK ultra-high vacuum scanning tunneling microscope for spin-resolved spectroscopy at high energy resolution, *Rev. Sci. Instrum.*, **75**, 4871–4879 (2004).

33. S. Meckler, M. Gyamfi, O. Pietzsch, and R. Wiesendanger, A low-temperature spin-polarized scanning tunneling microscope operating in a fully rotatable magnetic field, *Rev. Sci. Instrum.*, **80**, 023708 (2009).

34. K. K. Wang, W. Z. Lin, A. V. Chinchore, Y. H. Liu, and A. R. Smith, A modular designed ultra-high-vacuum spin-polarized scanning tunneling microscope with controllable magnetic fields for investigating epitaxial thin films, *Rev. Sci. Instrum.*, **82**, 053703 (2011).

35. C. G. Shull, W. A. Strauser, and E. O. Wollan, Neutron diffraction by paramagnetic and anti ferromagnetic substances, *Phys. Rev.* **83**, 333–345 (1951).

36. S. Foner, Versatile and sensitive vibrating-sample magnetometer, *Rev. Sci. Instrum.*, **30**, 548–557 (1959).

37. S. Blügel, D. Pescia, and P. H. Dederichs, Ferromagnetism versus antiferromagnetism of the Cr(001) surface, *Phys. Rev. B*, **39**, 1392–1394 (1989).

38. R. Ravlic, M. Bode, A. Kubetzka, and R. Wiesendanger, Correlation of dislocation and domain structure of Cr(001) investigated by spin-polarized scanning tunneling microscopy, *Phys. Rev. B*, **67**, 174411 (2003).
39. T. Kawagoe, Y. Iguchi, T. Miyamachi, A. Yamasaki, and S. Suga, Spiral terraces and spin frustration in layered antiferromagnetic Cr(001) films, *Phys. Rev. Lett.*, **95**, 207205 (2005).
40. A. Kubetzka, P. Ferriani, M. Bode, S. Heinze, G. Bihlmayer, K. von Bergmann, O. Pietzsch, S. Blügel, and R. Wiesendanger, Revealing antiferromagnetic order of the Fe monolayer on W(001): spin-polarized scanning tunneling microscopy and first-principles calculations, *Phys. Rev. Lett.*, **94**, 087204 (2005).
41. M. Takada, P. L. Gastelois, M. Przybylski, J. Kirschner, a complex magnetic structure of ultrathin Fe films on Rh (001) surfaces, *J. Magn. Magn. Mater.*, **329**, 95–100 (2013).
42. I. E. Dzialoshinskii, IEThermodynamic theory of weak ferromagnetism in antiferromagnetic substances, *Sov. Phys. JETP-USSR*, **5**, 1259–1272 (1957).
43. T. Moriya, Atomic superexchange interaction and weak ferromagnetism, *Phys. Rev.*, **120**, 91–98 (1960).
44. C. L. Gao, W. Wulfhekel, and J. Kirschner, Revealing the 120 degrees antiferromagnetic neel structure in real space: one monolayer Mn on Ag(111) *Phys. Rev. Lett.*, **101**, 267205 (2008).
45. M. Wasniowska, S. Schroder, P. Ferriani, and S. Heinze, Real space observation of spin frustration in Cr on a triangular lattice, *Phys. Rev. B*, **82**, 012402 (2010).
46. A. Wachowiak, J. Wiebe, M. Bode, O. Pietzsch, M. Morgenstern, and R. Wiesendanger, Direct observation of internal spin structure of magnetic vortex cores, *Science*, **298**, 577–580 (2002).
47. H. Oka, P. A. Ignatiev, S. Wedekind, G. Rodary, L. Niebergall, V. S. Stepanyuk, D. Sander, and J. Kirschner, Spin-dependent quantum interference within a single magnetic nanostructure, *Science*, **327**, 843 (2010).
48. O. Pietzsch, S. Okatov, A. Kubetzka, M. Bode, S. Heinze, A. Lichtenstein, and R. Wiesendanger, Spin-resolved electronic structure of nanoscale cobalt islands on Cu(111), *Phys. Rev. Lett.*, **96**, 237203 (2006).
49. R. Wiesendanger, I. V. Shvets, D. Bürgler, G. Tarrach, H. J. Güntherodt, J. M. D. Coey, and S. Gräser, Topographic and magnetic-sensitive scanning tunneling microscope study of magnetite, *Science*, **255**, 583 (1992).

50. C. L. Gao, A. Ernst, A. Winkelmann, J. Henk, W. Wulfhekel, P. Bruno, and J. Kirschner, Noncollinear surface spin density by surface reconstruction in the alloy NiMn, *Phys. Rev. Lett.*, **100**, 237203 (2008).
51. T. K. Yamada, R. Robles, E. Martinez, M. M. J. Bischoff, A. Vega, A. L. Vazquez de Parga, T. Mizoguchi, and H. van Kempen, Spin-polarized scanning tunneling microscopy/spectroscopy study of MnAu(001) thin films, *Phys. Rev. B*, **72**, 014410 (2005).
52. A. Leineweber, R. Niewa, H. Jacobs, and W. Kockelmann, The manganese nitrides η-Mn_3N_2 and θ-Mn_6N_{5+x}: nuclear and magnetic structures, *J. Mater. Chem.*, **10**, 2827–2834 (2000).
53. H. Q. Yang, H. Al-Brithen, A. R. Smith, J. A. Borchers, R. L. Cappelletti, and M. D. Vaudin, Structural and magnetic properties of η-phase manganese nitride films grown by molecular-beam epitaxy, *Appl. Phys. Lett.*, **78**, 3860–3862 (2001).
54. A. A. Khajetoorians, B. Chilian, J. Wiebe, S. Schuwalow, and R. Wiesendanger, Detecting excitation and magnetization of individual dopants in a semiconductor, *Nature*, **467**, 1084–1087 (2010).
55. A. A. Khajetoorians, J. Wiebe, B. Chilian, and R. Wiesendanger, Realizing all-spin-based logic operations atom by atom, *Science*, **332**, 1062–1064 (2011).
56. A. A. Khajetoorians, J. Wiebe, B. Chilian, S. Lounis, Samir, S. Bluegel, and R. Wiesendanger, Atom-by-atom engineering and magnetometry of tailored nanomagnets, *Nat. Phys.*, **8**, 497–503 (2012).
57. S. W. Hla, K. Braun, and K.-H. Rieder, Single atom manipulation mechanisms during a quantum corral construction, *Phys. Rev. B*, **67**, 201402 (2003).
58. J. A. Stroscio and R. J. Celotta, Controlling the dynamics of a single atom in lateral atom manipulation, *Science*, **306**, 242–247 (2004).
59. N. Atodiresei, J. Brede, P. Lazic', V. Caciuc, G. Hoffmann, R. Wiesendanger, and S. Blugel, Design of the local spin polarization at the organic-ferromagnetic interface, *Phys. Rev. Lett.*, **105**, 066601 (2010).
60. S. L. Kawahara, J. Lagoute, V. Repain, C. Chacon, Y. Girard, S. Rousset, A. Smogunov, and C. Barreteau, Large magnetoresistance through a single molecule due to a spin-split hybridized orbital, *Nano Lett.*, **12**, 4558–4563 (2012).

Index